Dr. Eleanor's
BOOK OF
Common Ants OF
California

DR. ELEANOR'S

BOOK OF

Common Ants

OF

CALIFORNIA

ELEANOR SPICER RICE,
ALEX WILD, *and* ROB DUNN

THE UNIVERSITY OF CHICAGO PRESS
Chicago and London

The University of Chicago Press, Chicago 60637
The University of Chicago Press, Ltd., London
© 2017 by Eleanor Spicer Rice, Alexander Wild, and Robert Dunn

Published 2017

Printed in Canada

26 25 24 23 22 21 20 19 18 17   1 2 3 4 5

ISBN-13: 978-0-226-35153-7 (paper)
ISBN-13: 978-0-226-39851-8 (e-book)
DOI: 10.7208/chicago/9780226398518.001.0001

LCCN: 2017026800

♾ This paper meets the requirements of ANSI/NISO Z39.48–1992
(Permanence of Paper).

CONTENTS

PREFACE

I grew up imagining I might someday go to a faraway tropical forest as an explorer. I imagined that in faraway places big new discoveries were still possible, discoveries of great and hidden empires. I was lucky enough to go to such places and even to make a discovery here and there. But as I have gotten older, I have discovered something even more fantastic. I have discovered that great and poorly explored empires can be found not just in the deepest jungles but also in backyards. New species and even whole societies remain to be studied in the dirt beneath our feet.

Among the least explored empires are those of the ants. Ants live nearly everywhere. They do not appear to have made it to outer space, but it seems only a matter of time. Some kinds of ants have been very well studied, but just as for beetles, mites, spiders, and other arthropods, most have not. That is why the tropical explorer Andrea Lucky and I, along with a large number of colleagues, started a project called the School of Ants.

In creating the School of Ants, we aimed to give kids and adults around the United States (and now in parts of Italy and Australia) the wherewithal to go into their backyards and collect ants in order to document where ants of different species live. The project is new, but already the discoveries have been big. One boy in Washington State discovered an ant species living in his backyard that was thought to live only in the southeastern United States, for example.

But knowing which species live where is just a starting point. Having found the empires of the ants, the real challenge is to spend the time necessary to learn their ways. The good news is that for each of the most common ant empires in backyards, major discoveries are still possible. This is what I wish I had known as a kid. I wish I had known that instead of (or simply before) heading away to tropical forests to make new discoveries, I could have made them in my own backyard. But there is a catch.

The catch is that in order to make discoveries, one needs to know what is already known, where the last path ended and where a new one might begin. Before now, there has been no book describing what we know about the common ants of North America. Most of what is written about the common ants describes how to kill them (which is a shame given that most common ants do no harm and offer a great deal of benefit to our yards and even our homes). But here, Dr. Eleanor, an ant biologist from Goldsboro, North Carolina, tells their stories. These stories are fun, but they are also something more; they are a clear indication of where the paths end in our understanding of these common species. Some of the ants Dr. Eleanor writes about are relatively well known (see, for example, the fire ant), but most are not, and even those that are well known await major discoveries.

I wish I had had *Dr. Eleanor's Guide* when I was young. I would have taken it—along with a bunch of glass jars, a shovel, a snake stick, and my other explorer's gear—out into the forest behind my house. With book in hand, I would have tried to add new information to the chapters. This is what I hope you do, because the truth is that each of Dr. Eleanor's funny stories about the most ordinary of our ants is just the beginning, and Dr. Eleanor needs your help to add new information. Perhaps you'll even help start a new chapter about a previously undiscovered species! And so go forth, young reader, and see what you can find.

Rob Dunn

INTRODUCTION

Before you dive into the stories of individual ant species, let's start with some basic ant biology and a little natural history.

Ants surround us, occupying nearly every type of habitable nook and cranny across the globe. Right now, ants snuggle up to your house, lay out their doormats in front of the trees in your yard, and snooze under park benches. Some even nest inside the acorns littering the ground!

We might not always notice them, but they're there, and they shape, literally shape, our world. Look at the colossal trees in our forests. Ants like winnow ants plant the forest understory, ultimately contouring the distribution of plants in the undergrowth and even of those giant trees. Other ants help turn soil (more than earthworms in some places!), break up decomposing wood and ani-

mals, and keep the tree canopy healthy by providing rich soil nutrients for saplings, protecting tender leaves from predators, and regulating tree pest populations.

Ants creep across our yards taking care of business for us in much the same way. They eat termites and chase caterpillars out of our gardens. Even though some people think of ants as the tiny creatures that ruin their picnics, of the nearly 1,000 ant species living in North America, fewer than 30 are true pests, and fewer still actually can hurt us. Most ants spend their time pulling threads, stitching together the quilt of the natural world. Without these threads, the quilt would fall apart, becoming disconnected pieces of fabric.

In this book, you will meet our most common ants. Odds are you can see these ladies tiptoeing all around you. See how beautiful they are, with their spines and ridges, their colors and proud legs, each feature lending itself to the individual's task. See their work, how they help build the world around us as they move about our lives.

What's in an Ant?

Like all insects, adult ants have three body segments: the head, the thorax, and the abdomen.

petiole

thorax

antennae

gaster

Heads Up

Windows to the world, ant heads are packed with everything ants need to interact with their environments. With tiny eyes for detecting light, color, and shadow; brains for memory and decisions; mouths for tasting; antennae for touching and smelling, ant heads are one-stop shops for sensory overload.

Thorax

Ant thoraxes are mainly for moving. Although an ant's nerve cord, esophagus, and main artery course through its thorax, connecting head to bottom, thoraxes are mostly legs and muscle. Every one of an ant's six legs sticks out of her thorax, and when queens and males have wings, those wings stick out of the top of the thorax.

Abdomen and Petiole: One Lump or Two?

The abdomen is where all the action happens. This ant segment holds all the critical organs. Almost all of an ant's digestive system is packed into its booty, as well as tons of chemical-emitting glands, stingers and trail markers, the entire reproductive system, and most of its stored fat. Many ant species also have a crop, which is a special stomach in their abdomens that does not digest food. Ants use this stomach as a backpack to carry food back to the nest. There they share the food with their sisters by vomiting it back up and then spitting it into their sisters' mouths.

The first part of an ant abdomen is called the petiole. The petiole is that really skinny section between the thorax and an ant's big fat bottom, which we often call a gaster. The petiole gives ants their skinny waists and flexibility when they move around. A lot of people interested in identifying ants check the petiole first to see if it has one bump or two as a first step in determining the species.

Where's the Nose?

Unlike us, ants don't have noses. Instead, they smell and breathe with different body parts. To smell, they mostly use their antennae. To breathe, ants have little holes all along their body called spiracles, which they can open and close. When the spiracles open, air rushes into beautiful silvery tubes that lace the ants' insides and bathe their organs with the oxygen they need to survive.

The Ant Life Cycle

Like butterflies, beetles, and flies, ants grow up in four stages: a tiny egg, a worm-like larva, a pupa, and the adult stage that most of us recognize as ants.

After hatching, larvae molt several times.

Eggs

For most species, only the queen lays eggs that become workers. Most of the time, eggs are creamy-colored orbs smaller than the period at the end of this sentence. When queens fertilize the eggs, females hatch. When they don't, males hatch. Sometimes, like when the colony is just getting started, queens lay eggs called trophic eggs. Delicious and nutritious, workers and developing larvae eat trophic eggs and use that energy to help the colony grow. Occasionally, workers lay eggs, but queens and other workers sniff about the nest and gobble those eggs as soon as they find them, no bacon needed.

Larvae

The only time an ant grows is when it's in the larval stage. Most ant species' larvae look like wrinkly grains of rice or chubby little maggots. Fat pearlescent white tubes with plump folds and wormy mouths, ant larvae are the chicken nuggets of the insect world. When colonies get assaulted by other

Workers tend eggs and larvae.

ants or insects, these little chub monsters are usually the first to go. Because they have no legs, they can't run away, and they make easy meals for anyone able to break into the nest.

Unable to feed themselves, larvae sit like baby birds with their little mouths open, begging for workers to spit food down their gullets. When workers give larvae special food at just the right time, their body chemistry changes and they grow up to be queens. As larvae grow, their skin gets tight on their bodies. They wiggle out of the tight skin as if shedding an old pair of blue jeans, revealing a brand new, bigger skin underneath. Never wasting the chance for a snack, workers squeeze whatever liquid remains from those discarded larvae skins and sometimes feed the solids back to larvae.

If you look at older larvae under a microscope, you'll see sparse hairs jutting out of their supple flesh. I know a scientist who wanted to find out why they have these luxurious locks, so he gave larvae haircuts and watched what happened. The verdict? Shorn larvae fall over like little drunk sailors. They need their hair to anchor them to surfaces.

Some ant species, like this one, spin silken cocoons.

Pupae

When larvae grow big enough, they quit eating and get really still. The larvae of some ant species will spin a silken cocoon around themselves to have a little privacy during this pupa stage; others will just let it all hang out. Although they don't look like they're doing much during this stage, their bodies are changing and shifting around inside that last larval skin. They're developing legs and body segments, antennae and new mouthparts. They're turning into the ants we recognize.

When they're ready, they squeeze out of their last skins, emerging as full-grown ants. At first, they tumble about like baby deer, unsure on their legs, soft and pale like the larvae they used to be. But after a while, their skins darken and harden, their step becomes surer, and they begin their work as adults.

Workers

The two most important things to remember about ant workers are the following: First, all workers are adult ants. Once an ant completes its metamorphosis (when it is recognizable as an ant), it will never grow again. When you see a little ant, it's not a baby ant; it's just a species of ant that is really small. Second, all workers are female. Workers do nearly every job, so pretty much every ant you see walking around is a girl. While queens get the colony rolling and keep it strong by laying eggs, workers get the groceries, keep intruders out, take out the trash, feed the babies, repair the house, and more. When we talk about ant behavior and the special characteristics of ants in this book, we're talking about the behaviors workers exhibit in the natural world, since they are the colony's only contact with the outside world.

Queens

Despite their regal moniker, ant queens are mostly just egg-laying machines. When queens first emerge, they usually have wings, but after they find a lucky someone (or someones) and mate, they rub off their wings, let their booties expand with developing eggs, and go to town eating food and popping out eggs. Protected deep within the nest, workers feed queens and keep life peachy for them so they can produce healthy eggs for the colony.

An ant queen, with her large gaster, is ready for egg laying.

Males

Male ants are easy to discount because they don't seem to do too much around the colony. Unlike their more industrious sisters, male ants refrain from cleaning up around the house, taking care of the babies, going out to get food, or keeping bad guys out. The one thing male ants do for the colony is mate with queens.

To date, scientists have spent very little time studying male ants. But these mysterious and weird-looking creatures invite a closer look. Compared to their sisters, most male ants have tiny heads and huge eyes. Often, they look like wasps. Nobody knows for sure what the boys do when they leave the nest. What are they eating? Where do they sleep? Why doesn't anybody seem to care? As you read this, researchers are trying to finally put an end to our male ant ignorance.

What's in an Ant Colony?

Many different types of ants will nest in pretty much any type of shelter. While fire ants push up their great earthen mounds for all to see, acrobat ants might have their mail delivered to a tiny piece of bark on a tree limb, and winter ants scurry down inconspicuous holes in the ground to their underworld mansions.

While ant nests differ greatly, when you crack one open, you'll most likely find lots of workers (the ants we most often see in the "real world"), a queen (many species have several queens), and a white pile of eggs and larvae.

Most ants carry out the trash and their dead, piling them in their own ant dumps/graveyards called middens. A good detective can learn many things from going through someone's trash. If you examine a midden, you can get a good idea of what the ants have been eating and whether or not the ants are sick or at war with other ants. You'll probably discover bits of seeds and insect head capsules stuffed in with the dead ants. When tremendous numbers of dead ants litter the piles, it's likely the colony is sick or warring with other ants.

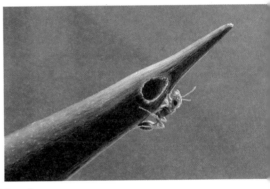

Ants nest in the ground, in thorns, and anywhere else.

Back inside the nest, the ants busy themselves with their daily anty lives. You can take some cookie crumbs and call them out to you. See how they sniff the earth with their antennae, each one a living being experiencing the world, doing its special job. Watch them communicate, following one another under blades of grass and around pebbles, stopping every now and again to touch one another's faces, clean their legs, investigate their surroundings.

Ants saturate our environment, from our homes to the sidewalks, city streets, and forests spread all around us. They are our neighbors, our friendly fellow citizens working away as we work. It's time we introduce ourselves.

01 PAVEMENT ANT

SPECIES NAME: *Tetramorium* sp.E (the ant formerly known as *T. caespitum*)

SIZE: 3.25–4mm (0.09–0.15in.)

WHERE IT LIVES: Pavement ants most often nest under bricks or pavement, but they are also found in grassy areas near sidewalks and even in extreme environments, like salt marshes. They live on the north, central, and south coasts, and, possibly, in the central valley and other regions.

WHAT IT EATS: Ultimate opportunists, pavement ants eat anything from dead insects to honeydew, a sugary food produced by sap-feeding insects like leaf hoppers. They also dine on pollen and food in your kitchen and garbage.

Wars happen across America every spring. Just as the trees begin to give us that first peek of color and the sun warms us enough to stretch our legs and venture outdoors for a look around, the animals begin stretching their legs, too.

Pavement ant battles are brutal. Here, four workers tear apart an ant from another nest.

Each spring, ants poke their antennae out of earthen holes, getting a feel for their new year on the beat. Workers of the pavement ant species (*Tetramorium* sp.E—although the pavement ant is common, scientists have yet to give this species a real name) push out of their nests with a mission: to establish their neighborhoods before ants from other nests nudge in and squeeze them out. These ladies are territorial, and they don't like any other ants walking on their turf. When they first emerge in spring, all the previous year's boundary lines have been wiped away with winter, and all bets are off. They redraw their property lines with warfare so gruesome it would make Attila the Hun blush.

Pavement ants are built for battle. At three-sixteenths of an inch, workers are about half as long as one of your shirt buttons is wide. They are dark reddish-black and have antennae that bulge out at the tips, making them look like they're waving little clubs from their foreheads. They have tough, armor-like skins called exoskeletons that can withstand the knocks of war. If a pavement ant were the size of a dog and you could get a good close-up look, you would see a beautiful landscape. Their faces and bodies are covered with hilly peaks, rivers of grooves and hairs, and they have two little mountains of spines poking out from their backs toward their rear ends.

Where neighborhoods overlap, huge numbers of workers from each side collide. They furiously drum one another's heads with

their antennae; they rip one another apart with their mandibles. They'll separate an individual from the pack and close in around her, gnashing at her body with their jaws, grabbing her with their claws, turning her into ant dust. These ants mean business when it comes to setting boundaries. After the melee, the carnage is astounding. Thousands of ants litter sidewalks across the country, a jumbled dark line of body parts and pieces that blow around in the wind.

When they aren't out cruisin' for a bruisin', pavement ants move along slowly compared to other ant species, as though they don't have anything to do in this big old world but go for a walk in nature. They won't sting you, and they aren't easily spooked. Whereas some ants shoo away quickly, pavement ants usually continue to bumble along unbothered.

Pavement ants are not native to the United States, but they are one of the most common species around. They sailed over here in ships from Europe more than 100 years ago and flourish in the stone-slab environments of modern cities. They most often build their nests under bricks and in sidewalk crevices and will eat everything from sugary foods to dead insects to flower pollen to human garbage.

Sometimes, pavement ants act like miniature farmers. They collect seeds from plants and accidentally plant them by burying them in their nests. They also tend insects called plant hoppers like dairy farmers tend cows, "milking" them for honeydew, a sugary food the plant hoppers produce. If a plant hopper predator comes lurking around, pavement ants will pick the plant hoppers up in their mouths and carry them down to their nests, where they'll wait out the trouble. Pavement ants also keep interlopers off their property and will wipe out any ant nests that try to pop up on the homestead. But this is all during peace time.

Back to spring. The birds are practicing their songs, and you and I are hopping off the school bus, picking up lucky pennies, walking our dogs, or going to get coffee on our sidewalks that zig and zag from New York City down to Florida, across Tennessee, the Dakotas, and Wyoming, all the way to California. Each day, as we walk around in our world, the human world of sidewalks that point us to and from where we want to go, we are also walking over the world of the pavement ant, with devastating wars, property disputes, and peace times filled with farming and baby making. Their world is so similar to ours, so close to us, that we step over it every day without noticing how unusual these ants are.

02 WINTER ANT

SPECIES NAME: *Prenolepis imparis*

A.K.A.: false honey ant

SIZE: 3–4 mm (0.12–0.15 in.)

WHERE IT LIVES: Winter ants nest deep in the soil near tree bases or in open ground, like lawns. They live on the north and central coasts, in the Sierra and Coastal Mountains, and in the central valley.

WHAT IT EATS: While winter ants won't pass up an opportunity for a sugary snack, these ladies prefer protein-packed food, noshing on other insects not lucky enough to endure winter's chill.

Remember in *Alice in Wonderland* when Alice followed the white rabbit down its bunny hole? The hole was ordinary enough at first, but once Alice climbed in, she fell down and down until she came to a completely different world. Holes like that rabbit's pepper the

ground across the United States. If we were as small as ants, we could tumble repeatedly down into other worlds. Winter ants are the white rabbits of ants. Plunging down their holes gives us a peek into their truly extraordinary lives.

Unless you follow a winter ant home, its nest's entrance can be hard to find. About the size of a buttonhole, winter ant nests aren't a lot to look at on the outside. Inside, however, deep mazes of tunnels connect chambers all the way to the bottom. The nests can extend almost 12 feet deep into the soil. That would be the human equivalent of a class of second-graders digging a hole more than 1.14 miles down, deep enough that 150 school busses could be stacked end-on-end before reaching the surface.

All that depth serves a purpose. While most ants are active in the spring and summer, winter ants prefer the fall, winter, and spring. Soil temperature does not vary as wildly as the temperature above ground, so when winter's chill plummets to 33°F, the winter ant's nest is kept insulated by the earth and remains at a balmy 64–68°F. This heat is important because between 40–50°F, most insects develop a serious case of brain freeze, going into what bug people call a "chill coma," where their muscles stop working so they can't move. Underground, winter ants beat the ice. Above ground, they dig short "warming tunnels" scattered around their nest. When they start to get cold walking around outside, they run down into the tunnels and warm up.

Staying Out of Trouble

My mother always told me the best way to stay out of trouble is to avoid it. Winter ants are masters at avoiding trouble because they move about when trouble is fast asleep. From March to November, when most ant species scramble around gathering food and fighting one another for space, winter ants seal themselves tightly in their nests. When November rolls around and other ant species tuck themselves in for their winter naps, winter ants unseal their nests and begin exploring the world.

Because they are active when other ants sleep, they often miss the dangerous tides of invasive ants that can wipe out many other ant species. In this way, they can persist in areas inhabited by other inhospitable ants. If they do happen to meet an adversary, they spray a toxic chemical from their rumps that scares off or even kills the would-be contender.

How to Spot Them

At the beginning of winter, winter ants are hard to identify. Shiny reddish-brown with lighter yellow legs, they look like your everyday, run-of-the-mill ant. Early in the season, workers are about 0.1 inch long, just long enough to span the letter *t* on this page. But as the season progresses and winter ants stock up on food, they become easier to identify.

To say a winter ant has a big behind mid- and late season is an understatement. When workers eat their favorite protein foods like insects and a sugary substance produced by other insects called honeydew, they stockpile the calories in special fat cells in their bums. These fat cells can grow to be of tremendous size. Because they wad-

Sometimes called "false honeypot ants" for their large rear ends, winter ants use that extra junk in their trunks to survive summer underground.

dle around with swollen rumps at the end of the season, some people call winter ants "false honeypot ants."

To understand why they pack on the pounds, let's poke our heads down into one of their rabbit holes. Winter ant colonies survive underground all summer on their rotund sisters' fat. Their ample behinds are the world's best refrigerators. Because fat cells are part of living tissue, as long as the worker is alive, the fat won't rot, like dead insects stored in the nest would. And because the fat is already concentrated and high in calories, workers don't have to process it like they would other foods. Winter ants store enough fat in their portly posteriors to feed each other and all the babies that emerge as adults the following fall. When workers unseal their nests in the fall, they emerge as Skinny Minnies again.

Home Deep Home

Let's travel a little deeper down the rabbit hole. While its worker inhabitants live a couple of years at most, winter ant nests can exist for more than ten years. The older the nest, the deeper it is. If we were winter ants crawling down into our home, we would enter through a short hallway leading to the first room. Other than the pinhole of light shining through the entrance, the whole house would be completely dark. To get from room to room, we'd have to smell our way with our antenna. Our rooms would have domed ceilings, tall enough for a couple of us to stand on top of one another. Because we'd have clingy feet, we could even walk on the ceiling!

We might have a few hundred sisters—sometimes up to 10,000—living with us, so every now and again, we'd bump into one of our sisters and give her a friendly tap with our antennae. If she seemed hungry, we might spit up a bit of food for her to eat. If she seemed dirty, we'd help clean her with our mouths and antennae.

It might take us a long time to get all the way to the bottom of the nest. Remember, winter ant nests are at least the human equivalent of a mile deep. Our older sisters would live in the upstairs rooms, and our younger sisters would live with our mothers deep down. Our queen mothers would wander around the bottom of our nest in the dark laying their eggs. Our younger sisters would help feed the babies and keep them clean, while our older sisters would gather food for us.

Life Underground

If we were winter ants, we would not be able to hear well, and it's quiet so far underground anyway. We wouldn't hear children running over us or leaves falling on our entrance. We wouldn't know somebody's dad's car just parked next to our own driveway. Beneath the roots, we wouldn't get wet when the sprinkler showers over our home and across the lawn in the summertime. We wouldn't hear the thud of the family dog flopping right on top of us to gnaw on a tennis ball. But it would all be happening above us, all over the United States. If we were winter ants, we'd miss out on a lot of the fascinating lives of people. We're lucky we're not winter ants. We're people, active all year long, and able to understand and delight in the winter ant's secret wonderland, deep below our feet.

03 THIEF ANT

SPECIES NAME: *Solenopsis molesta*

SIZE: 1.5–1.8 mm (0.06–0.07 in.)

WHERE IT LIVES: North, central, and south coasts, central valley, Sierra and Coastal Mountains. Thief ants nest underground in forests and open, grassy areas. They also like to nest in human structures. They particularly like nesting near other ant species' nests.

WHAT IT EATS: Tiny ants with big appetites, thief ants prefer protein, such as dead insects.

Back in the days of the Wild West, Jesse James and his outlaw gang were some pretty crafty dudes. They robbed everything from stagecoaches and trains to banks and homes. His bandit bunch crept into towns and would hightail it out ahead of angry lawmen and WANTED

posters bearing James clan faces. Imagine if the Jesse James family moved in right next door to your house! Many ant species across the United States face this predicament every day when thief ants come to town. Thief ants are the Jesse James gang of the ant world, and these bite-size burglars pickpocket and plunder anything they can get their little mandibles around, living lives of artifice that would make Mr. James sit up and take some notes.

Even though he was a robber and a murderer, Jesse James won the public's hearts, in part because he was easy on the eyes. Thief ants are no different. Whenever I stumble upon a thief ant nest or happen to lift a dead insect and find a bevy of thief ants, mid-snack, I always stifle a squeal. Thief ants are unbelievably, ridiculously cute.

Their size might play a big factor in their cuteness. At one-sixteenth of an inch, a thief ant worker could wander comfortably around in a lower case *o* on this page. Most often a golden yellow color, thief ant workers vary along the color spectrum all the way to amber. They have stingers, but they are too tiny to cause you any pain. They look like they wander around really slowly, but actually they're just super small. If you had a microscope, you could see that each antenna has a bulb on its end, and they bonk about as they feel their way to and from food. Much of that food, remember, is stolen, either from other ants or from you and me.

Thief ants get their name from their habit of setting up camp next to other ant species' nests. They love protein and stuff their bellies on dead insects, people food, and insect eggs. When the other ants bring home thief ants' favorite foods, those crafty little burglars sneak that food right on over to their own houses and feast. They've also been known to smuggle out other ants' babies, tasty snacks for greedy thief ants. When other species' colonies are weak or dying, thief ants aren't as sneaky. They run through the nests' halls like children running down the aisles of a Toys"R"Us on a shopping spree, eating their fill of dead and dying ants.

Their crimes and misdemeanors don't stop with the insect world: A thief ant will rob you blind if you don't watch out. Thief ants are opportunists, and they recognize that your kitchen is a wonderful opportunity for the biggest heist of their lives. Because they are so small, many people have a hard time figuring out how to keep them out of their pantries. The best way to keep thief ants out is to figure out how they're getting in. Once you do that, block their entrance-way by plugging holes with some caulk or weather stripping and tell those thief ants there's a new sheriff in town.

Some people think Jesse James was like a modern-day Robin Hood and that many of his crimes were to benefit others. I don't know what Jesse did with all of his loot, but many of the thief ants' crimes against other insects surely do help us out a lot. For example, when they're not stealing from other ants, they love to eat lawn pests like cutworms and scarab beetle eggs, and they provide effective control against these lawn and golf course pests.

Even though they're miniscule, they're pretty good at bullying one of our biggest ant bullies: the red imported fire ant. Thief ants are almost three times smaller than the smallest fire ant. Like the James gang, they rely on their cunning and strength in numbers to beat up and eat any upstart fire ant colony making camp in their territory. In fact, fire ants can't establish nests in areas where thief ants roam.

Being tiny has its advantages. Because thief ants nest underground and out of sight, they are one of the few ant species who can weather the havoc wreaked by other nasty invaders like Argentine

ants and yellow crazy ants. When other ant species get kicked out of town, thief ants hold their ground.

Jesse James's shoot-'em-ups and looting sprees came to an abrupt end when he met the wrong end of Robert Ford's pistol. Fortunately, thief ants survive even the toughest ant assassins. They beat up fire ants and outwit Argentine and yellow crazy ants. Unlike Jesse, who caused trouble with

Thief ants tend root aphids underground.

the law wherever he went, thief ants contribute to our natural world. They help keep other pieces of nature in check by eating dead insects and aerating the soil with their underground nests. You could even say they are lawmen in their own right, nibbling away at the pests crawling around your lawn. They're tiny but tough, and they're outside your door right now. Despite their name, thief ants live mostly on the good side of the law.

04 ODOROUS HOUSE ANT

SPECIES NAME: *Tapinoma sessile*

SIZE: 2.25–3.2 mm (0.09–0.13 in.)

WHERE IT LIVES: Odorous house ants nest indoors (under sinks and doormats and in insulation and dishwashers) and outdoors (under rocks and in garbage cans, potted plants and exposed soil). They are found along the north, central, and south coasts, in the Central Valley, in the Sierra and Coastal Mountains, and in the desert.

WHAT IT EATS: Odorous house ants eat honeydew, a sugary liquid made by small, sap-feeding insects like aphids and scales, and other sugary food left out by humans. They also eat dead insects and spiders.

People across the United States call me all the time to tell me they have ants in their houses. It's one of my favorite parts of knowing a little bit about insects. From my grandmother Ina down in Opelika, Alabama, to my good friend Ariana out in Los Angeles to my friend Sarah's grandmother's friend up in Baltimore, the call is always the same: "Help me! I'm under attack! I've got ants in my kitchen!"

I love these calls because they make me feel like a real live wizard. Here's why: Across the United States, there are only three or four types of ants that often wander into people's kitchens. By asking a few questions, I can usually narrow the identity of the particular trespassers down to the species through a process of elimination. It's simple, but it seems like magic to the people who are calling. To let you in on the secrets of my sorcery, here is the phone conversation I had with Sarah's grandmother's friend (SGF):

SGF: "Help me! I'm under attack! I've got ants in my kitchen!"
ME: "Are they big or little?"
SGF: "They're tiny!"

Clue 1: They are tiny. Now I know she doesn't have big carpenter ants or the less probable field ants. She also doesn't have Asian needle ants.

ME: "What color are they?"
SGF: "I gotta look at them? Hold on. I gotta get my reading glasses. Hold . . . on . . . OK! They're black!"

Clue 2: They are black. So, Sarah's grandmother's friend doesn't have pharaoh ants or fire ants. Plus, she probably doesn't have the brown Argentine ants. One more answer and I'll know what she has in her kitchen. Time for my big finish.

ME: "Here's what I want you to do. I want you to squish one. I want you to roll it between your fingers and put it up to your nose and sniff it."
SGF: "I'm sorry, what?"
ME: "Just do it. Tell me what it smells like."

Sarah's grandmother's friend squishes. She makes the I'm-squishing-an-ant sound people make, which comes out as a mix between "ooh!"

(fun!) and "eew" (gross). The result of this squish-and-sniff will tell me whether she has little black ants (about half the size of a sesame seed) or odorous house ants (a little bigger than a sesame seed).

Clincher: They have an odor. Like most people with ants, Sarah's grandmother's friend has odorous house ants partying in her kitchen. Their telltale smell gives them away. She's a lucky lady. Neither dirty nor dangerous, this top home pest—also known as the sugar ant—can provide hours of entertainment for anyone willing to share space with them. Follow them home to see how they bunk! Put out food and see how long it takes them to find it! Lay an *E.T.*-style trail of snacks to shift their ant highways! Possibilities for fun abound.

Country Ant, City Ant

Unlike some of the ant species that pester people around the country (imported fire ants or Argentine ants, for example), odorous house ants did not migrate here. They are US natives. Named for a defensive odor they emit from their rumps that some describe as

"spoiled coconut suntan lotion," they nest in natural environments like the woods or in pretty much any manmade locale like potted plants, under doormats, or in cars. As with Aesop's country mouse and city mouse, "country" odorous house ants (those living in natural, wooded areas) and "city" odorous house ants (those living in manmade environments) lead different lifestyles.

In the country, odorous house ants play an important role keeping the earth a clean, green machine. They work in concert with other forest bugs to keep tree canopies healthy and ensure a proper ecological balance with plenty of species hanging around. They also help accelerate decomposition and promote nutrient flow by eating dead insects and animals and nesting in and under rotting wood, in acorns, and in abandoned insect homes.

Yes, out in the country, they live the quiet life and have small colonies of a few hundred to a couple thousand workers. But once they move into cities, odorous house ants go a little wild. Their populations explode, sometimes spanning entire city blocks, and they blanket lawns and kitchen counters with greedy scouts sniffing around for a sugar fix.

When we build cities, we also build the perfect environment for odorous house ants to go berserk. First, it's easy for them to find a job to help support their city lifestyle. Plenty of ant employers looking for work (aka scale insects and aphids) await in the trees we plant to line our neighborhood streets. These creatures depend on odorous house ants to protect them from ladybugs, tiny wasps, and lacewings, all aphid and scale predators. When odorous house ants show up, those predators split, enabling aphid and scale populations to soar. To pay for their security detail, aphids and scale insects provide odorous house ants with a sweet syrup called honeydew.

In the woods, odorous house ants compete with different species for places to set up camp. With acorn ants stuffing their homes into acorns, citrus ants pouring out from under tree bark, and acrobat ants peeking down from tree branches, odorous house ants make do

Odorous house ants milk aphids like cattle for a sweet honeydew reward.

wherever they can. But in the city, they can nest anywhere. Vacancies abound. From our garbage cans packed with odorous house ant–ready foods to the luxurious mulch we pile up around our homes to our kitchen floors, odorous house ants feast, raise babies, and have shindigs around us all the time. City odorous house ants can have many nests per colony with tiny superhighways of workers moving between them, distributing supplies from nest to nest. Some odorous house ant colonies can span a city block.

In the country, as conditions around their nests change, such as when a new, more dominant species comes to town or a big storm floods the area, odorous house ants move out. They generally move their nests every two weeks or so. This ability to pack up and move willy-nilly in the woods helps them cope with ever-shifting, human-made environments. Garbage day? Dumpster-living ants can saunter over to the grassy area. Dumping out those potted plants? Odorous house ants who had been living inside happily toddle over to the compost pile. Having many queens in the nest helps them split up without too many tearful goodbyes.

Odorous house ants carry larvae around the nest in their mandibles.

Roll Up the Welcome Mat

While I see odorous house ants in my kitchen as a happy surprise, I'm aware that not everybody (OK, probably not most people) shares my sentiment. It can be disconcerting to see eager sugarbears trundling across your Wheaties. After I conduct my wizardly identification, the response never seems to be: "What FUN!" It's almost always: "How can I get rid of them?"

Store shelves are packed with poisons designed to extinguish these ladies. However, knowing what we know now about odorous house ants, most of us can outsmart them. Be a detective. Stake them out. Follow them home to see how they are sneaking into your house. Then, eliminate the access point. We know that odorous house ants like to hang out in tree canopies and bushes, slurping up honeydew. Walk around your house and see if you have any bushes touching your walls or windows. Branches bridge the ants from their outdoor lifestyles to apartment living. Cut back those branches. Snoop out other ways they enter the house. For example, they sometimes sneak in through cracks and crevices. Seal those with caulk.

We know they love to nest in mulch. People often dump piles of mulch around their homes. Switch that out for rocks, which odorous house ants don't like as much. Or try aromatic cedar mulch, which smells gross to odorous house ants, at least for a little while.

Look where they're crawling around inside, too. We know odorous house ants like sugar and all the delicious little treasures abundant in human garbage, so don't leave food out and tightly seal garbage cans. But even if you try to get rid of these sweethearts, pay attention as you do. Because the truth is, most of what might be known about these ants hasn't been uncovered. Most of their tiny empire's treasures lay undiscovered. So, while I can tell you as much as I've told you about sugar ants, I can't tell you much more. When someone calls to tell me about their sugar ants, most of what they have to report is not just grievance, it is science.

And so, when Ina says, "They keep stopping and talking to each other with their antennae," or Ariana reports, "I left my Coke open and they found it in less than 30 minutes!" these are things I write down, things you might want to write down too.

05 CARPENTER ANT

SPECIES NAME: *Camponotus semitestaceus, C. vicinus*, and *C. modoc*

SIZE: 7–13 mm (0.25–0.51 in.)

WHERE IT LIVES: Many carpenter ants prefer to nest in living, standing trees or under stones but will also nest in logs and wood in human structures. *Camponotus modoc* lives in the Sierra and Coastal Mountains, along the North and Central Coasts, and in the Central Valley. *C. semitestaceus* lives along the North, Central, and South Coasts, in the Central Valley, and in the Sierra and Coastal Mountains. *C. vicinus* lives along the North, Central, and South Coasts, in the Central Valley, in the desert, and in the Sierra and Coastal Mountains.

WHAT IT EATS: Omnivores, carpenter ants eat protein foods, including other insects, as well as sugary foods, like honeydew produced by aphids.

Carpenter ants make excellent ambassadors between ants and humans because of their size—they are among the United States' largest ants—and often pleasant disposition.

Between a quarter-inch and a little over a half-inch long, a small carpenter ant can comfortably stretch over a plain M&M, and a large one can just about reach across dime. Unlike some workers in some other ant species, carpenter ant workers vary in size and shape within the colony. You might find big girls with fat, tough heads wandering trails alongside their scrawnier, nimbler sisters. They might look different, but they're all sisters, and they're all full-grown adults. Colonies have between about 350 to several thousand workers patrolling a variety of environments, from street curbs to forest floors. In addition to noting its large size, you can identify a carpenter ant by the light dusting of golden hairs on its head and thorax that settle on its abdomen.

Three carpenter ant species make California's most common list: the night-loving *Camponotus semitestaceus*, the tree-chomping *C. vicinus*, and the beautiful-but-fighty *C. modoc*. The former two species can look very much alike. *C. semitestaceus* have rusty reddish-brown to yellowish-brown thoraxes with darker heads and abdomens. *C. vicinus*, usually darker than their *C. semitestaceus* cousins, have deeper red thoraxes and darker blackish heads and abdomens. *C. modoc* are dull black all over with beautiful reddish legs. They more often live in the Northern Californian mountains but can be found farther south.

Carpenter ants nest under stones or in soil, building little mounds around their nest entrances. All three species have those golden-haired booties, and all three species offer hours of fun for anyone interested in ant watching.

Breakfast for Ants

When I was little, I took my breakfast crumbs out to my front yard to feed the carpenter ants living in the willow oak trees. I would build little piles of bacon and toast for them on top of oak leaves and wait for them to lumber out from holes hidden in the bark at the base of the trees. They were my connection to the ant world. Later, I learned that carpenter ants peer out from tree bases across the United States.

I loved those ants. I was fascinated by the way they walked around like miniature black horses, exploring their way with their elbowed

antennae, stopping every now and then to gently tap their sisters and give each other waxy kisses. If I pressed my ear against the tree near their entryway, I could hear them crackling about their business inside. If I sat still, they would come up to my hands and gingerly pick crumbs off my fingers. If I picked one up, she would explore my arm and shirt. If I squeezed her, she would give me a pinch with her tiny jaws. It never hurt.

They're called carpenter ants because they are particularly good at woodworking. *C. modoc* and *C. vicinus* like to nest in living, standing trees using their sturdy mandibles to excavate tunnels and rooms in dead limbs or in dead wood at the tree's center. They don't actually *eat* the wood; they just whittle out rooms for their homes. You may see carpenter ants living in your white maple or fir trees and think the ants are killing the trees. But carpenter ants actually have a history of helping trees. They have an appetite for tree pests like red oak borers, and they spend a lot of their time foraging around their home, plucking pests off the bark. The trees housing my carpenter ants 25 years ago are still standing today.

Because of these woodworking skills, some people see carpenter ants as household nuisances. While some carpenter ant species, like *C. modoc*, can carve passageways in the wood of people's homes, they often point homeowners to bigger problems: damp and rotting wood from a leak or drip, or other pests living in that wood. When wood becomes soaked through, carpenter ants can easily use their jaws to snap it away and bore their tunnels. If homeowners keep their wood dry, carpenter ants will usually stick to the trees. That is, unless the homeowners have pests like termites or wood beetles snacking away inside their walls. These pests lure in carpenter ants, who pounce on this abundant food opportunity like me at an all-you-can-eat buffet. One can hardly blame them! If you see a carpenter ant in your house, follow her to her home to see if you can find out why she's there. She might lead you to a bigger problem. Not all carpenter ant species want to live in your home. *C. vicinus* may break inside peoples' homes to forage, but they prefer their soil nests and stick to the outdoors most of the time. Shy *C. semitestaceus* also usually keep it natural outside.

Ant Speak: Decoded

Unlike many Golden Staters, carpenter ants try to avoid basking in the famous California sun. If you shine a light on foraging *C. semitestaceus* workers, they will hide from it under blades of grass or by running down their nest entrances like little vampires hiding from the sun. I used to think my carpenter ants might like some of my bologna sandwiches from lunch, but I couldn't get as many takers at lunchtime as I got early in the morning. That's because carpenter ants are mostly night owls, foraging from dusk until dawn. With their big eyes, carpenter ants have pretty good vision for ants, and they use that vision to help them take shortcuts from their house to food in the early morning and when the moon is out.

When they aren't following their sisters' chemical trails, they remember landmarks like pebbles and sticks to help them find their

way home. These landmarks save time for carpenter ants, who can sometimes forage up to 100 yards from their nest. That's the human equivalent of walking more than 11 miles for food. On new moon nights, when it is totally dark, carpenter ants take no shortcuts and feel their way through the night, keeping their bodies close to structures.

When carpenter ants find food, they run back to the nest, laying a chemical trail behind them. Once inside the nest, they do an "I found something awesome" dance to get their sisters awake and excited enough to follow them. The hungrier the ants, the more vigorous the dance. The excited sisters then rush out of the nest in search of the chemical trail that will lead them to the food. Carpenter ants, like many other ant species, have little built-in knapsacks called crops inside their bodies. They fill these crops with liquid food to take back home. When they meet their sisters on the trail, they stop and have a little conversation that goes something like this:

ANT HEADING OUT TO FOOD: "Hey, what's up?"

ANT RETURNING FROM FOOD: "Are we from the same nest?" (They check this by tapping each other on the head with their antennae to see if they smell alike.)

HEADED-OUT ANT: "Yeah, but I'm not sure what I'm even doing here. I'm just following this trail." (She moves her tapping antennae closer to her sister's mouth.)

RETURNING ANT: "Oh, wow! I should have told you earlier. Some kid spilled his Dr. Pepper down the street, and it is DELICIOUS. Everybody's over there now drinking it up. Want to try?"

HEADED-OUT ANT: "That sounds awesome. Of course."

Returning ant spits a little droplet from her crop into headed-out ant's mouth. Headed-out ant drinks it and agrees it is awesome. Awesome enough, in fact, to continue running down the trail.

When I was a child, I saw carpenter ants having these sorts of conversations all the time and thought they were kissing. When I grew up, I learned that I already knew much about carpenter ants from watching them as a child. Their colony size, where they nest, and how they eat have all been scientifically dissected and explored as thoroughly as the ants themselves explore the dark tunnels of their homes. Scientific papers explain how they talk to each other, when they're awake, and why they don't want bologna on hot summer afternoons. Every delicate golden hair on the carpenter ant's rump has been counted and cataloged. These discoveries took many decades to document. All of them can be remade any morning by each one of us, holding our breakfast crumbs, waiting patiently in our front yards.

06 SOUTHERN FIRE ANT

SPECIES NAME: *Solenopsis xyloni*

SIZE: 2.5–4.6 mm (0.1–0.18 in.)

WHERE IT LIVES: Southern fire ants nest in mounds or flattened craters in open soil near moisture. They also can nest under carpets, in crawl spaces, or under rocks. They live in the Central Valley, along the North, Central, and South Coasts, and in the Sierra and Coastal Mountains.

WHAT IT EATS: Southern fire ants have a healthy appetite for pretty much anything, including dead insects, sweets, greasy foods, and sometimes seeds.

Here is a story about a villain, a stinging sensation, and a possible hero. It is the story of the Southern fire ant, known to scientists as *Solenopsis xyloni*. Easily confused in appearance and behavior with the red imported fire ant (*Solenopsis invicta*, that notoriously bad guy

Southern fire ants make quick work of a tasty grasshopper.

from the Southeast), Southern fire ants generally make their (much smaller) homes in small, loose dirt mounds in grassy openings or under rocks or boards in the southwestern United States.

Our story begins with a villain and a few simple facts about the birds and the bees. Actually, just one fact: birds and bees dread Southern fire ants.

As ground-dwelling predators, Southern fire ants can devastate ground-nesting birds like bobwhite quail. Although a quail adult outsizes a Southern fire ant worker 160,000-fold, Southern fire ant nests can have upward of 15,000 workers, who comb the ground with their pinchy mandibles and venomous stingers, swarming quail nests and devouring their much smaller chicks. In fact, Southern fire ants rank up there with coyotes, skunks, and badgers as one of the top four predators of bobwhite quail nestlings in Texas. But their bad behavior toward birds doesn't end with bobwhites. Southern fire ants also attack house sparrow chicks and endangered least tern babies nesting on California's beaches. Killing Southern fire ant nests increases least tern populations because the ants aren't around to chomp away at their fuzzy little babies.

While the gruesome death of one of nature's most squeezably fluffy inventions (baby birds) may seem horrible enough, plants might argue they've got it even worse.

Let's be a plant for a minute, maybe a happy little bush. Here we are in our field, and life is pretty sweet. The sun is shining on our leaves, and our flowers are just beginning to open and say hello to the world. We smell so good, enticing fat little bees to come bury their fuzzy heads deep within our petals for a drink of nectar. Hello, bees! As they burrow in for a sip, they dust their heads with our pollen and transfer it to other flowers. We need these bees to visit us so we can do the one thing we need to do most in this big old world: reproduce. Make seeds.

Normally, we love ants. They keep the number of pests squiggling around on us stable so we don't get sick or depleted. But to us, these Southern fire ants are the *worst*. First, instead of controlling

pests, they protect insects that can give us disease or suck us dry, harboring them against predators in exchange for a nectary treat the pests give them as a reward.

Lots of other ant species do the same thing, but not many are as good at it as Southern fire ants. Southern fire ants are such fierce protectors of their flocks of pestilence that they scare away anyone else who comes near us, including pollinators like bees and butterflies.

To make matters worse, these ladies rob nectar from our flowers. That is, instead of dusting their bodies with pollen and moving that pollen from flower to flower like bees do, these little jerks bypass our pollen when they drink up our nectar. That's right. Not only do they make us lousy with all those pests, they also prevent us from making seeds, our only task in life!

By now you might be thinking, these ants are just terrible. But I know one ant expert named Andrea who would disagree with you. "I love Southern fire ants," she told me once. "They're so cute and shy. They always try to run away from you. Besides, they're native and they're always getting pushed around by those other fire ants."

This is the other side of Southern fire ants, the side that holds a valuable place in nature. They evolved with the natural environment across the southern United States, helping to regulate the balance of animals and plants in their natural home. Southern fire ants used to live all across the southern United States, but around the 1950s, red imported fire ants and their big, grumpy colonies began trouping across the Southeast, stomping out many Southern fire ant nests in their way.

Have you ever met somebody with a bad reputation and expected to dislike them, only to find out that they were actually pretty nice? As Andrea pointed out with her observation, this may be the case with the Southern fire ant; it's possible that we just haven't really *met* them yet.

With science, we only know what we have studied so far. We've studied Southern fire ants' bird-eating, plant-hurting behavior, so that's their reputation for now. But we have a lot left to learn about these insects. How has the environment changed since red imported fire ants came and Southern fire ants left? How do Southern fire ants behave around other ant species? How often are we mistaking the deeds of red imported fire ants for those of Southern fire ants? I'm sure you could come up with some good questions about Southern fire ants yourself.

We can build on and challenge what we know through exploration. You and I can form our own opinions based on information. Isn't that wonderful? With science, you and I can each meet these ants, ask questions about them, and discover for ourselves more chapters in the Southern fire ant's story.

07 FIELD ANT

~~~~~~

SPECIES NAMES: *Formica integroides* and *F. moki*

SIZE: 5–10.2 mm (0.2–0.4 in.)

WHERE IT LIVES: Field ants generally build their nests against trees, under rocks, or in logs. *Formica integroides* are found on the North and Central Coasts and in the Central Valley. *Formica moki* are found on the North, Central, and South Coasts, in the Central Valley, in the desert, and in the Sierra and Coastal Mountains.

WHAT IT EATS: More buffet goers than picky eaters, field ants love sugar such as aphid honeydew, soft-bodied insects like caterpillars, and seed husks.

~~~~~~

Formica ants, usually called field ants, are among the United States' largest and most common ants. With many species found spanning the states in all directions, two make California's most common list: *Formica integroides* and *Formica moki*, both beauties with orange-

to-deep-red heads and thoraxes and darker abdomens. Most field ants pass their days contentedly building their shallow, low-mound nests near rocks and trees, blissfully unaware of a dark underworld in their midst, a world of violence, slavery, mistaken identity, and poop shields.

About the size of one and a half pencil erasers, field ants' long, dexterous legs extend from their thoraxes. Their large, black eyes rest right behind their always-moving elbowed antennae. Many people confuse field ants with carpenter ants, neither of which can hurt you. If you'd like to tell if you have a field ant, gently nab the ant in question and check out its thorax, the middle section of the ant where all the legs attach. If its thorax consists of two lumps, you have a field ant. If it has one big hump, you're holding a carpenter ant. What a way to impress your friends!

Field ants have large eyes because they usually move around during the

Field ants milk aphids for sweet honeydew.

day and rely on sight more than some other ant species. They use those big eyes to help them see landmarks as they scurry to and from food. Like many ant species, field ants love tending aphids and scale insects for their sugary emissions, but they also help disperse plants by toting seeds around the forest, snacking on the husks and discarding the rest. They enjoy wolfing down other insects whenever they get the chance. Sometimes, *Formica moki* will wander into your home looking for food.

Nice Outfit, Mr. Beetle

Field ants prefer to eat soft-bodied insects like caterpillars and beetle larvae, and this predatory tendency helps keep our trees happy. One of the United States' most dangerous forest pests is the gypsy moth. Thanks to their huge appetites, gypsy moth caterpillars have gobbled up more than 80 million acres of our northeastern forests in the past 40 years. When they scarf down all the leaves in the forest, trees die, causing millions of dollars' worth of damage. Fortunately, field ants love those plump little leaf munchers. They help reduce the damage and spread of gypsy moths by eating every caterpillar they can find.

Sometimes their partiality for pudgy little insects lands field ants in unusual situations. Many baby beetles (often called grubs) fit the mold for a perfect field ant meal. Slow, soft, and chubby, beetle grubs don't stand a chance when hungry field ants stumble across them while foraging. To ward off potential beetle slayers, many beetle species, like tortoise beetles, rely on an inventive solution: poop shields.

Here's how it works: Some plants in our forests and across our cities have certain "stop eating me!" chemicals in their leaves, called deterrents. When most insects bite into a leaf and smell the deterrents, they get as far away as possible. Not our resourceful beetle grubs. They eat as much of these stinky leaves as they can, pooping stinky leaf poop all over the place. Then they gather up the poop and stick it on their bodies, making a force field of stink that follows them wherever they go. Field ants catching a whiff of these otherwise tasty tidbits run in the opposite direction of our little Pigpens. If you feed these baby beetles nonstinky plants, they still make a poop force field, but because it contains no deterrents, field ants will ignore the BM blanket and eat them right on up.

Slaving Away

It may seem like all fun and games for field ants, frolicking across our forests, lawns, and traffic medians, grocery shopping and building their houses. But field ants have a wicked foe prowling those same forests, lawns, and traffic medians, combing the grass for field ant nests: Amazon ants. Amazon ants look a lot like field ants: same size, similar color, same big eyes, similar camel-humpy back.

Amazon ants carry slaves, unlucky field ant pupae.

Amazon ants and field ants look so similar they could pass as the same species, almost: Amazon ants have dagger-sharp, sickle-shaped jaws. Their jaws are so pointy they can't take care of tender babies—any attempt at carrying or feeding could result in a fatal stab wound to their young.

So Amazon ants came up with a solution: They raid field ant nests, frighten adults into submission with poofs of chemicals, snatch up hearty pupae in those jaws, and scurry back to their nests. Now, we remember from the ant's life cycle that baby ants take a lot of food, but once those ants pupate, they don't eat at all. They just sit there helpless in their nests and wait to turn into adults. By stealing pupae, Amazon ants basically snatch up soon-to-be adult workers that require no maintenance in the meantime.

Once in the raiders' nest, the field ant pupae start to pick up the smells in the nest. Ants tell one another apart by smell. If a field ant starts to smell like an Amazon ant, she'll start to think of herself as an Amazon ant. When she emerges as an adult, she will do the tasks to help the colony that she would have done in her real mother's nest: gathering food, building the nest, raising babies, taking care of the queen. She usually will have no idea that she's a slave, helping her enemies to grow so they can raid more field ant nests.

Each summer, poor field ants are enslaved up and down the United States, from the forest near my North Carolina house to the parks of busy Long Island, New York. But you and I can still spot those lucky enough to have escaped the dagger-jaws of the Amazon ants. They run along our tree trunks and across our sidewalks, planting seeds, snagging bugs, turning soil. We can look for their double humps and drop them a snack like a piece of a cookie or some juice from our juice boxes and see if they eat it. We can give them advice, telling them to stay away from poison poop and to keep their big eyes peeled for slave makers. They can give us advice, too. They can tell us never to underestimate the power of small things, to be mindful of the good they can do. They can tell us that every animal has a complicated story, a life of adventure and trials that unfolds whether or not we humans pay attention. But you and I can pay attention. Field ants are happy to share their story with us.

08 HIGH NOON ANT

SPECIES NAME: *Forelius pruinosus*

SIZE: 1.8–2.54 mm (0.07–0.1 in.)

WHERE IT LIVES: Masters of climate, high noon ants can nest just as happily in your kitchen cabinet as they can in the middle of the desert. They prefer grassy or open ground and often nest under rocks in the central valley, the north, central, and south coasts, and the desert regions of California.

WHAT IT EATS: Excellent scavengers, high noon ants will eat meaty foods like dead insects and animals, but they often prefer liquid sweet treats, like those produced by aphids.

The first time I ran into a high noon ant worker, I'll admit I was underwhelmed. I was ant hunting in a grassy park, laying bait to see which ant species lived in this anty jungle. I'd brought along the

perfect enticement: tuna fish mixed with honey. I measured out this ant catnip onto an index card in a tiny spoonful, which I placed on a spot of bare ground under an oak tree. Then I lay in wait to see who would show up.

Before long, many of my old friends came nosing around. A rusty red field ant with speedy long legs was the first to arrive at the party, bending down, legs spread wide like a horse, to drink in the buffet. She was followed by a small flock of odorous house ants, who were chased away by a steady throng of shiny little black ants. A few acrobat ants briefly lurked around the index card's borders considering the feast then returning to their tree, evidently thinking better of it.

As the little black ants began to scatter, a collection of tentative ant workers I didn't recognize loitered in a tidy line on the sidelines. Plain Jane—brownish red and about half the size of an apple seed— these ladies were unremarkable in appearance. Unlike the frantic field ants or the spirited little black ants, they were a bit boring.

Watching these austere, drab ladies as they efficiently carried off the remaining bits of index card bounty, I almost felt sorry for them. Where were the great spines of winnow ants? The gargantuan size of wood ants? The giant noodles of big headed ants? The happy, heart-shaped bottoms of acrobat ants? Unembellished at best, high noon ants (very common ants who only got a common name as a result of the School of Ants study) don't make a knockout first impression.

As I became better acquainted with them, I learned there's more to high noon ants than meets the eye. So, in salute to the lesser known (and maybe less appreciated) high noon ant, I give to you a countdown of what I believe to be her top five, most notable attributes:

Five. High noon ants are masters of climate. They love to nest in open areas and are able to survive just as well in temperate fields and your kitchen or bathroom as they can in deserts.

Four. They smell good. Like odorous house ants, high noon ants have a pleasant aroma when you smash them. Their bottoms are packed with a chemical that smells sweet, almost like something you'd use to clean your counters. But don't be fooled by their bouquet bottoms: This smell-good chemical is actually an alarm pheromone that attracts their nestmates, who then form a mob to help their sister in danger.

High noons eat meaty insects and sweet honeydew.

Three. These clever ants have an entrepreneurial spirit. They have figured out a way to trade their bodyguard skills, whether they're looking out for other insects or for plants, for their favorite syrupy treats. Take the catalpa tree, which gets the most out of this ant's fondness for dessert. When catalpa trees put out their leaves, hungry little Jabba the Huts called catalpa caterpillars come to gobble them up. If nobody protects the tree from these fat piggies, catalpa caterpillars can eat every last leaf off the tree, hurting the tree's ability to produce the food it needs to survive. To save themselves from starvation, catalpa trees put out an SOS call, to high noon ants.

As soon as a caterpillar chomps down on a catalpa leaf, the tree oozes nectar along its branches, which attracts high noon ants. The ants don't want anybody messing with their sugar stash, which means the unsuspecting caterpillars are out of luck. Other plants, like wild cotton plants, similarly benefit from high noon ants' fierce love of sugar.

But high noon ants have no special loyalty to plants. When caterpillars offer to pay them in sugary treats, high noon ants are quick to take the job. Sometimes, they visit and protect the endangered Miami blue butterfly caterpillars, who also offer them a sweet reward for their efforts.

Two. They have the absolute *worst* party etiquette. To keep other diners off the buffet, high noon ants go to those ants' nest entrances in large numbers. Instead of dropping off an invite, they spray bug repellent on the nest entrance, driving the nest's inhabitants deep down. Then, they block up the entrances so those other ants can't escape. Voila! All you can eat is ready and waiting just for the high noon ant.

One. High noon ants can dance! Whenever these ladies are faced with conflict or danger, they try to appease their opponents with a little jig. They shake their bodies around like they're doing the jitterbug and then point their bottoms in each other's faces. This is how they size each other up. Performing these dance moves might not make us the belles of our balls, but they save high noon ants from a lot of cuts and bruises.

So what if they don't have giant spines or huge heads? So what if they don't have cool-shaped bodies? As it turns out, while high noon ants may *look* polite and buttoned up, they're anything but run-of-the-mill workers. With their sweets-loving, booty-shaking, party-crashing ways, high noon ants are a great reminder that we should never judge an ant at first glance.

09 ARGENTINE ANT

SPECIES NAME: *Linepithema humile*

SIZE: 2.2–2.6 mm (0.09–0.1 in.)

WHERE IT LIVES: Argentine ants prefer cities and form loose nests in moist landscaping materials such as mulch, under trees or walkway bricks, in potting soil, or under rocks. In California, they are found on the North, Central, and South Coasts and in the Central Valley.

WHAT IT EATS: Argentine ants scavenge and devour insects and other small arthropods and human garbage, but they prefer sweet liquids such as the honeydew produced by scales and aphids.

When I first started studying ants, my friend and lab mate Alexei took me to an office park just outside of Raleigh to hunt for ants. Seeing nothing but boring, run-of-the-mill brick office buildings sep-

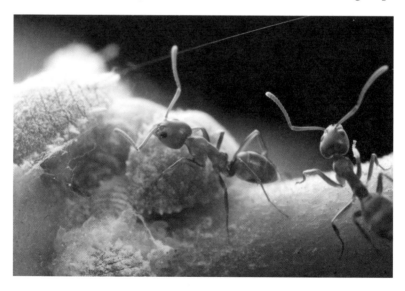

arated by expansive parking lots dotted with sickly trees, I started to think he had misunderstood our mission. Where were the forests with acrobat ants waving their fannies happily along tree trunks? The lush backyards of strangers hiding thief ants, winnow ants, and maybe a fat, grumpy box turtle or two? At the very least, where were the odorous house ant–packed dumpsters behind city restaurants— restaurants that might have an ice cream cone waiting inside just for me? Alexei pulled into one of the parking lots. It looked like all the other parking lots in the park.

"We're here," he said, and we hopped out of the truck.

We wove between parked cars and skipped over empty spaces. He peered behind sedans and pickup trucks, crept around coupes. Meanwhile, I developed a working theory about the strength of his sanity.

Suddenly, Alexei pointed. "Look right there!"

Hugging the curb, a thick braid of dead ants stretched the length of the parking lot as far as I could see in either direction. It twisted and bent in the wind.

Alexei explained we were standing on an Argentine ant (*Linepithema humile*) supercolony boundary line. He said they fight to defend their colony borders every day, which accounts for the huge number of casualties. I picked up a piece of the dead ant rope and looked more closely at these Argentine ants. About the size of two sesame seeds and reddish brown, they didn't look like much. Soon I learned that, despite their looks, I was holding a fistful of some of the most extraordinary ants in the world.

Argentine ants set sail for the United States from Argentina and Brazil in the late 1800s. Stowaways in ships' ballasts, they made port in New Orleans and quickly set up camp. Before long, they fanned out and blanketed the city, nesting in every nook and cranny they could find. They gobbled up everything they could. One researcher in the early 1900s wrote that Argentine ant populations were so humongous, they ate all the bedbugs in town.

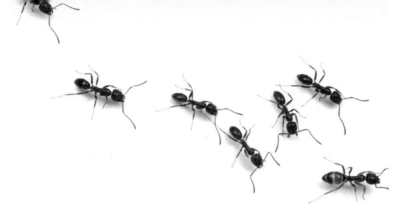

Back at home, while their rambling sisters were out gobbling up all the crawfish étouffée they could eat, the native Argentine ants weren't a problem, quietly nesting in floodplains and other moist areas. Today in their native land, they still aren't a problem. Argentine ants, like many other ant species, can tell the difference between ants from their home nest and ants from other nests based on the way they smell. When Argentine ants encounter an ant that smells differently from the way their home nest smells, they attack it. As a result, Argentine ants keep each other in check and their colony sizes stay small.

When Argentine ants hitch a ride to other countries, only a few families make it to establish a home. These families, closely related to each other, smell a lot alike, so the workers refrain from fuss-

This Argentine ant queen will work with other mothers in a super-colony to lay thousands of eggs.

ing and fighting. Instead, they concentrate their energy on fighting other ant species, gathering food, growing... and growing. By keeping the peace with each other and dominating other species, Argentine ant colonies turn into supercolonies. While the supercolonies in my state of North Carolina stretch only a few miles, in California one Argentine ant supercolony spans more than 500 miles. Along the Mediterranean coast, one reaches almost 4,000 miles. These supercolonies contain thousands of nests and queens with millions of workers.

All this buddy-buddy behavior among invading Argentine ants can really wreck their adopted lands. Taken alone, one Argentine ant isn't all that impressive. With no spines or hairs, no impressive size or fancy gasters, they look like regular, run-of-the-mill ants. They can't sting. Squishing an Argentine ant is no problem—they can't even run that fast. When individual Argentine ant workers try to defend themselves against other ants, they rarely stand a chance.

Argentines use teamwork to overcome larger enemies.

But with millions of sisters getting each other's backs, they can team up and overwhelm any opponent. They handily defeat native ants and scarf down all kinds of insects, mites, and other arthropods, and their appetite for sugar leads them to foster sugar-producing plant pests like aphids and scale insects. Altering the environment on such a dramatic scale can have cascading effects across the whole ecosystem.

When Argentine ants remove other ant species, they take away food sources for some animals, and they eliminate the beneficial work of other ants without filling those positions. For example, by kicking out ants native to California, Argentine ants eliminated coastal horned lizard food, leaving baby lizards to starve and lizard populations to plummet. In South Africa, they give seed-dispersing ants the boot, and the plants that grow from ant-dispersed seeds die out. By protecting honeydew-producing (and also disease-transmitting) plant pests, Argentine ants increase plant diseases in their invaded areas. Their sweet tooth also drives them into honey-bee hives in parts of the country, bothering the bees so much that they abscond for ant-free areas, leaving beekeepers to shake their fists. Left unattended, Argentine ants can empty a whole beeyard of hives.

With so many scouts combing the area for snacks, Argentine ants can locate ant food faster than other ant species. Once at the snack site, they get all their sisters involved to bully any other would-be snackers into staying away.

Protecting the environment from these little raiders can be extremely hard. They have picked up a few tricks from their native land that keep them just ahead of the insect control curve. Because they evolved in floodplain areas, where at any time their home could be underwater, Argentine ants are good at moving from place

Argentine ants gather honeydew from scale insects.

to place. Their many-queened nests are loose groups of ants under mulch, doormats, pine straw, in home walls—anywhere moist and dark—with a cluster of white brood and eggs. When danger strikes, they pack up the young 'uns and move. As they move, they can split their queens into more nests. Now, one nest has become several, and the Argentine ant population has room to grow.

Because of this, people often find Argentine ants breaking into their homes and office buildings, scarfing down their sodas and hanging around their trash cans. We try to knock them back from

our property, unaware that millions of them wait on our borders, ready to move into the empty spots left by their sisters.

Yes, tiny Argentine ants cause all sorts of trouble. Yes, it would probably be best if we sent them all packing back to Argentina, where their natural enemies and cousins would give them a stern talkin' to. But for now, Argentine ants aren't going anywhere. Some researchers are capitalizing on this, watching how Argentine ants run around when they're panicking, to help us understand how people can safely exit dangerous, crowded situations.

After that first day with Alexei in the office park parking lot, watching the dead Argentine ants blow like a long tumbleweed in the wind, I spent many years observing Argentine ants. I watched their brown bodies glimmer in the sun as they trailed up and down maple tree trunks, carrying honeydew in their fat bottoms. I saw pieces of popcorn seem to float across the pavement as these ants carried them back to their nest. I watched them disassemble a whole grasshopper before my eyes. I've uncovered nests all thick with white brood and eggs and watched them carry that whole nest to a new location in minutes. I saw these tiny individuals persist, grow, and dominate by working peacefully together.

10 *POGONOMYRMEX* HARVESTER ANT

SPECIES NAME: *Pogonomyrmex californicus* and *P. rugosus*

A.K.A.: California ant and rough harvester ant

SIZE: 5.6–8.9 mm (0.22–0.35 in.)

Where it lives: Harvester ants build nests in sandy, hot areas. *P. californicus* lives in the desert, in the Central Valley, and along the South Coast. *P. rugosus* lives along the South Coast and in the desert.

WHAT IT EATS: Although they will scarf down arthropods like grasshoppers and caterpillars, they prefer to gather seeds.

If you're ever in the desert, you might notice flat earthen disks spotting the landscape. They look like the footprint of a recently fled UFO, with the surrounding vegetation cleared and pushed to the out-

side of the circle, and sometimes flecks of ancient treasures—fossils, charcoal, parched leaves—spread across their surfaces. If you step up to one, instead of the left-behind rubbish of alien invaders, you'll see a single hole in the center. And if you're there in the afternoon when the sun shines its brightest, you can watch harvester ants bubbling out of that hole and pouring outward across the earth. The simple act of harvester ants going about their business transforms the world around them and can seem almost otherworldly.

Harvester ants love warm, sandy areas across the United States, but they especially love the West, and many species make their home there. Two of California's most common species are *P. californicus*, the California harvester ant, and *P. rugosus*, the rough harvester ant. Both species range in size from about the length of a popcorn kernel to about the length of a staple, and both species are good-looking ants.

With long, narrow waists and lanky legs, these lookers can dress up any old desert floor. But they move less delicately than their bodies suggest. Their squarish heads pointed forward, they charge ahead with a determined military crawl. Up close, both species of

Pogonomyrmex are covered in intricate wrinkles and folds, making them look as if they've been left in the bath a little too long.

California harvester ants, the ants famous for being shipped to

hopeful young myrmecologists across the globe as residents of "Uncle Milton's Ant Farms," are a bright, rusty red with beautiful black eyes and a gaster that ranges from the same rusty red as her other two segments to a deep mottled black. Much darker overall, the color of rough harvester ants varies from a dark reddish-black to a deep black-brown. These deep hues are not, ahem, slimming, and so despite being about the same size, rough harvester ants seem stockier than California harvester ants.

Harvester ants have luxurious beards.

Both species are speckled with stiff, white hairs across their bodies and boast long, very unladylike beards under their chins. These beards, called psammophores by ant scientists, give harvester ants their genus name, which means "bearded ant," and they also help them to do their job of carrying sand around when doing nest chores. The beards act as baskets, gripping otherwise slippery sand as the ants excavate and clean.

When harvester ants excavate their nests, they often pull tiny fossilized mammal bones from underground and scatter them across their nest area. This consequence comes in handy for fossil hunters in the western United States, who save themselves the trouble of digging at random by first inspecting harvester ant discs for bones to see if it's worth checking out the area for fossils.

Although harvester ants eat insects like cicadas, carob moths, and lygus bugs, foraging for seeds is their primary occupation. They get to work in early afternoon and continue to forage until midafternoon. Older workers scour the ground and climb plants for seeds.

If they find something too big to carry back, they dip their abdomens on the ground and lay a trail with chemicals from their poison glands—also meant for stinging and rated one of the most potent toxins in the world. They bring seeds back to the nest, where younger workers process them and cache them for later in underground chambers.

Unlike dead insects, seeds don't rot quickly, so they store well. In unpredictable landscapes like deserts, where food might come easy after a rain or might not come for a long time, ant silos filled with food come in handy. But seed storage doesn't just help the ants. Harvester ants' occupation reaps benefits for the earth around them as well.

Harvester ant colonies can stick around for many years. Sandy soils like those harvester ants call home can be famously barren of nutrients and the tiny microorganisms that help process those nutrients. Harvester ants, in picking up seeds, plant parts, and insects and bringing them back to their nests make oases of nutrients, plants, and microorganisms. These fertile lands in the middle of the desert have more nutrients, like nitrogen and phosphorous. Harvesters also excavate the ground, and, in doing so, aerate it, which aids in providing an ideal environment for microorganisms to live in.

Harvester ants help decompose carrion by eating it, and dead plants and animals scattered near harvester ant nests decompose faster than they do when they expire out of the ants' reach. More plants thrive in harvester ant nest areas than anywhere else around the desert. And after a drought or a fire, those harvester ant–planted and –protected seeds sprout, helping the landscape to recover more quickly. By gathering and planting seeds and excavating their nests, harvester ants shape the entire desert landscape.

Harvester ants usually send their newly born queens and males out into the world in the summer, after it rains. At this early stage, the queens have wings, so they can fly out to find a mate. Most ant species—and most species of harvester ants (including rough harvester ants)—that have queens who found nests will do so alone. But sometimes California harvester ants help a sister out.

Sometimes, two young California harvester ant queens work together to found their new colony. One queen digs the nest and gathers up snacks to keep them fueled while the other queen lays eggs and takes care of the new babies. After the first workers emerge, both queens start to act like "normal" ant queens and turn into egg-laying machines, while their workers take care of the housework. The more queens that help out, the better odds the colony has of surviving, and that's a big deal because usually less than 1 percent of harvester ant queen hopefuls survive to make a nest.

Because they work together, these queens can start out with extra rooms for grain storage, deeper nests, and food to eat while they lay their first batch of eggs, so they won't have to use up all their precious fat supply. As a result, they have more energy to raise babies than queens who work alone.

Far from being creatures that aliens left behind, as their nests suggest, harvester ants are an important part of our planet. But they can seem alien to us—unlike anything we've known. These tiny creatures do the work of giants, fostering life where little life can grow, propelling gargantuan cycles across tremendous landscapes with little more than their golden beards, their spindly legs, and their ability to work together.

11 VELVETY TREE ANT

~~~~~~

SPECIES NAME: *Liometopum occidentale*

A.K.A.: velvety tree ant

SIZE: workers: 2.5–6 mm (0.1–0.24 in.)

WHERE IT LIVES: Velvety tree ants like to nest under tree bark or in the crevices of hardwood trees like oak, elm, elder, or cottonwood. Sometimes they'll move into your house's cracks and crevices. They live on the North, Central, and South Coast, in the Central Valley, in the Sierra and Coastal Mountains, and in the desert.

WHAT IT EATS: Velvety tree ants are omnivores; they'll eat anything from honeydew—a sugary liquid made by small insects like aphids—to scales and some caterpillars, to dead insects and human garbage.

~~~~~~

For a period of about three years, from ages 11 to 13, there was no mistaking me coming down the street. During that time, I discovered the most glorious material on earth: velvet. After a brief-but-luxurious encounter with a hair ribbon, I decided that velvet was the fanciest and most necessary material on earth. Despite gentle urgings from family and friends to consider more practical apparel, I began to envelop myself in all manner of velvet any chance I got.

I paraded the halls of the Wayne County, North Carolina, school system like 16th century royalty: creamy, velvet dresses for school dances two years running; soft velvet bows and ribbons falling into the mud at soccer practice; velvet leggings pumping down Mulberry Street after my dog, Sam; even a splendid multicolored velvet vest that jazzed up every ensemble—particularly my crushed velvet bodysuit. For some reason still unknown to me, the trend never caught on with my friends. In fact, they still remind me today of my fantastic fashion sense whenever I try to help them pick out clothes.

Upon our introduction, California's velvety tree ant scurried right into my heart with its own version of my favorite fabric. But though my velvet wardrobe set me apart from everyone else, people often confuse the velvety tree ant with other species.

Velvety tree ants squeeze into hard-wood crevices and tight spots in for-ests and coastal California homes. Between a tenth of an inch and almost a quarter-inch large, adults can be anywhere from about the size of an odorous house ant to a small carpen-ter ant. They have golden-to-reddish-orange thoraxes sandwiched between dark brown heads and gasters. As their name suggests, their heads and gas-ters are covered with short, light hairs that give them the matte shimmer of velvet, making them look like a tribe of little Henry VIIIs regally bustling along the bark.

Velvety tree ants have the deliber-ate gait of carpenter ants and prefer to nest in and around trees, so people often confuse the smaller, furrier ants with their larger carpenter ant cous-ins. Velvety tree ants also spray a defensive odor like odorous house ants, so folks finding them on their kitchen counters may confuse these two species. But just one close look at the velvety tree ant can set you back on the right track. And that's much the better for you, because knowing you have velvety tree ants can give you an after-noon's worth of ant adventures.

Tiny hairs give velvety tree ants a matte shimmer.

When it comes to family, velvety tree ants go big or go home. Each colony can have between 40,000 and 60,000 workers. They can have many nests and connect those nests with a network of hallways 1–2 cm deep, dug into the ground beneath the leaf litter. They use these hallways to reach other nests or to forage for food like sweet hon-eydew or insects. Their underground highway system keeps them cool and moist, and though they forage outdoors in the daytime in spring and early summer, as the California sun heats their stomping grounds, they forage underground more often, particularly at

night. Some of their foraging trails can be as long as 185 meters (607 feet). At human scale, that would be the equivalent of a tunnel 45 miles long—nearly the length of San Francisco.

Beneath their classy couture, velvety tree ants hide a scrappy attitude and are easily disturbed by would-be aphid eaters and humans alike. When bothered by humans, they run around, waving their pointy little mandibles in the air and emitting an alarm pheromone. These mandibles aren't just for show; they can pinch—a defense mechanism that becomes particularly bothersome if you happen upon a group.

Velvety tree ants sound the alarm.

Sometimes, velvety tree ants use the limbs that brush against your house or apartment as little bridges to wiggle their way inside, nesting in between walls or in other damp cavities. With their scrappy attitude and heavy appetites, you may walk into your kitchen one day to find yourself the keynote speaker at a convention of tiny, well-dressed creatures. You can help keep them in their proper home by cutting limbs back from your outside walls or by caulking crevices and holes.

Despite my enthusiasm and persistence, the abundant velvet look never came into style. And much to the delight of people who walked around in public with me, eventually I abandoned it, returning to my usual nonvelvet jeans and T-shirts. But velvet ants are a timeless classic. California natives, they have been classing up the understory for thousands of years, enhancing their environment by turning the soil for their tunnels, gobbling up extra insects, recycling nutrients, and looking swank while they do it.

12 *VEROMESSOR HARVESTER ANT*

SPECIES NAME: *Veromessor andrei* and *V. pergandei*

A.K.A.: black harvester ant and desert harvester ant

SIZE: 4.5–7 mm (0.18–0.28 in.)

WHERE IT LIVES: *Veromessor* harvester ants prefer living in very hot areas. You can identify a *Veromessor* harvester ant mound by the pile of seed husks on top. Often, you'll find a greater diversity of plants around their mounds, an unintentional garden caused by the germination of some of the seeds the ants drag back into the nutrient-rich and warm soil resulting from the bits and pieces of waste the ants throw away around their nests. *V. andrei* live along the North, Central, and South Coasts, in the Central Valley, and in the desert. *V. pergandei* live in the desert, along the South Coast, and in the Central Valley.

WHAT IT EATS: While *Veromessor* harvester ants prefer seeds, they also scavenge on arthropods.

A *Veromessor* harvester ant carries a seed.

High on the hill overlooking the city of Duckburg, magnate Scrooge McDuck's money bin sits chock-full of gold and treasures, unspent, ready for Scrooge to come luxuriate in his riches. *Veromessor* harvester ants are the Scrooge McDucks of the desert, tycoons of Ant-town, with their own money bins stockpiled with riches below their earthen mounds.

Veromessor harvester ants thrive in the hottest parts of North America. Whereas many animals, including other ants, prefer lush forests or busy cities packed with food almost year-round, harvester ants can survive more than a decade of severe drought conditions. More than 100 *Veromessor* species troll through grasslands and deserts, but two top California's most common list: *V. andrei*, the black harvester ant, and *V. pergandei*, the desert harvester ant.

Black harvester ants aren't actually black. They're more of a deep red, almost the color of ripening mulberries. With their spiky, white hairs from antennae to gaster's tip and their fingerprint-like impressions on their exoskeletons, a sort of inscrutable evolutionary tattoo, they look both rough and furry all at once.

Desert harvester ants, on the other hand, are black, smooth, and shiny. With constellations of short, white hairs springing from their burnished bodies, they look like little patent leather shoes in need of a good shine.

If you ever see a harvester ant and wonder whether it's a *Veromessor*, just follow her home. You'll know a *Veromessor* residence by its front-door décor: *Veromessor* harvester ants decorate their mounds with seed husks from their favorite meals.

Many plants grow around a *Veromessor* mound.

Often plants abound around *Veromessor* mounds because the heat and humidity in and around ant nests (resulting from the ants' breathing and also their wastes) promote healthy soil life that, in turn, fosters hearty plant growth. As seed specialists, *Veromessor* ants are forever dragging home seeds, and so many are retrieved that some end up around the entrance to the house. Whether by good luck or some other mechanism, some seeds avoid being eaten and sprout up, circling the nest entrance in lush plants. Because the soil around *Veromessor* nests is so healthy, some plants produce more than twice as many seeds when they grow around the ants' mounds than they do when they're left to struggle in the poor desert soil. The number of plants around a *Veromessor* mound can be five times greater than anywhere else around.

Those plants' seeds, the gold of the desert, are the secret to *Veromessor* ants' success and their Scrooge McDuck ways. In harsh environments, it can be tough to get a steady meal. Meaty insects are scarce and, much like a leftover piece of pizza that somehow makes it way under the couch, don't store well in the heat. But seeds! Seeds can be nutritious and keep fresh for a long time. Stored seeds can be food-rich during times when the environment offers no food. *Veromessor* ants know how to get rich.

Each morning just before dawn, *Veromessor* ants send out tens of thousands of workers from their nest entrance to travel in lines like the spokes on a wheel, looking for food. These foraging lines can be more than 130 feet long—the equivalent of our walking almost nine miles out to find something to eat then hauling it nine miles back. Over time, these foraging trails rotate around the nest, ensuring the ants cover the territory around them. When it's really dry outside, *Veromessor* ants, who need to stay hydrated, don't waste any time. They hustle as quickly as they can, making fast grabs for seeds and running back to the nest. Other days, they take their time, laying chemical trails from their abdomens when they find something tasty, to encourage their sisters to follow them.

In addition to their sisters' chemical trails, *Veromessor* ants use the sun to help them find their way home. They keep mental tabs on the angle of the sun in the sky compared to their nest location and time of day. Some scientists like to play tricks on *Veromessor* ants. They use a shade and mirror to make ants think the sun is shining from the opposite direction than it actually shines. When the scientists do this trick, *Veromessor* ants notice the sun's position, confidently turn around in the opposite direction, and walk away from their nest. When they remove the mirror and shade, *Veromessor*

ants see the sun's actual location, turn right back around, and head straight to their house.

Veromessor ants can be choosy when it comes to picking up their loot. They look for areas with as many seeds as possible; once they hit the jackpot, a single ant might examine more than 60 seeds before finding just the right one to carry home. And with what seems whimsy (they may have a reason scientists have yet to understand), *Veromessor* workers heartily pick up seeds from one plant one day but totally ignoring that same plant the next day. Once they find the right seeds, workers neatly pile them below ground in nest chambers, their money bins, holding on to them for tough times and protecting them from other seed eaters like desert-roaming mice. With some seeds, *Veromessor* ants eat the tasty, nutritious outer shell called the elaiosome, while with others, they eat the whole seed, adding its husk to their outdoor nest decorations.

Their frugal seed squirreling pays off; *Veromessor* ants are the dominant species in many desert grasslands, coasting through tough times that could scorch or starve other species by availing themselves of their well-stocked (and chosen) larder.

There's a lesson in saving from these Scrooge McDucks of the desert—a philosophy that's worked for thousands of years. Making runs in the desert sun, maintaining one's compass, and nurturing fertile ground in order to allow others, be they young sisters or lazy brothers, to grow pays dividends during tough times.

DON'T GET CONFUSED!

While closely related to their seed-loving cousins, winnow ants, *Veromessor* harvester ants are more often confused with the other desert harvesters, *Pogonomyrmex*. Both kinds of harvester ants have graceful, spindly legs and squarish heads with big eyes, but *Veromessor* are smaller with rounder bottoms, and, if you look closely, have a dip in their thoraxes that cause them to look like little hunchbacks scurrying across the sand.

Veromessor

Pogonomyrmex

Since both species share the same habitat and eat the same sorts of foods, it might seem that they'd be prone to fighting, but they avoid conflict by foraging at different times of day. While *Pogonomyrmex* like to come out in the middle of the day, *Veromessor* prefer making a run for it, once at dawn and again at dusk. *Pogonomyrmex* run into *Veromessor* ants only in passing when their shifts change.

13 HONEYPOT ANT

SPECIES NAME: *Myrmecocystus testaceus*

A.K.A.: honeypot ant

SIZE: 3–4.6 mm (0.12–0.18 in.)

WHERE IT LIVES: Hot weather lovers, honeypot ants usually stick to hot, arid areas across the western United States and Mexico. They dig deep nests in open areas. You can tell a honeypot ant nest by its large opening circled by a crater of soil. They live along the North, Central, and South Coasts, in the Central Valley, and in the Sierra and Coastal Mountains.

WHAT IT EATS: Honeypot ants love sugar and scour ground and plants for any type of syrupy liquids they can find, particularly from honeydew-producing, sap-sucking insects such as aphids and scales.

On the day after Thanksgiving, my mother and I engage in a ritual shared with millions of other Americans that makes almost no sense. It's called Black Friday, the day when stores mark down several limited-supply items to unbelievably low prices and open their doors before dawn to entice customers to buy their holiday gifts. Here's what happens in our house.

A little before four o'clock in the morning, my mother wakes me up to go to one of my hometown's few giant stores. Armed with a flyer from the previous day's paper announcing what will be on Major Markdown sale, my mother and I spend a few minutes in the parking lot trying to go over our game plan for getting in this store and getting out without losing each other or getting beaten up by strangers. Usually, before we decide how our reconnaissance mission will work, one of us will stop and say, "Who *are* these people?" as we look at the mayhem already unfolding at the store's entrance.

People of all ages, shapes, and sizes crowd in front of the doors. These people are what we call professional doorbusters, people who understand the ins and out of Black Friday. Then, the doors open. People go crazy. They push each other aside; the huge crowd squeezes through the tiny doors a few folks at a time. Once inside, it's a mad-grab for everything in the store. Toys, televisions, cardigans, coffee cans, pencil packs—everybody snatches and fights for anything they can get to mark off their holiday list. True doorbusters, in it to win it, can grab everything on their lists before others even make it through the store's entrance. Then it's off to the next store. Honeypot ants are the doorbusters of the desert.

Honeypot ants have long, spindly legs.

About the size of the caraway seeds in your rye bread, honeypot ants have long, rabbit ear–like antennae and large black eyes that stand out against their

Back in the nest, some workers called repletes dangle from the ceiling, storing nectar in their crops.

orange-golden color. Their long faces with blunt mandibles lend them a sort of buck-toothed appearance. Short, fuzzy hairs cover their legs and bodies. Their nests look like soil-ringed craters with big holes (big for ants, anyway) in the middle. Beneath those craters, honeypot ants toil away in rooms and chambers extending deep into the soil all across the western United States' hot, dry earth. Their big front doors and deep nests help honeypot ants survive their scorching environment.

To beat the heat, every day just before sunset, our doorbusting honeypot ants assemble at their nest's entrance. They crowd their big, buck-toothed rabbit faces at the entranceway, jostling and bumping each other, getting ready like those Black Friday bargain hunters. All the way down their entrance tunnel, the ants line up, squeezing to get out. As soon as the sun goes down, they take off, running across the sand, going gangbusters to get as much food as they possibly can in this cooler part of the day.

They run to nearby plants in a manic hunt for sugar, looking for aphids and scale insects that suck the juices out of plants and turn

those juices into syrupy honeydew for ants to gather. They find extrafloral nectaries—nectar oozing from odd spots on plants—and slurp it up as quickly as they can. While the doorbuster honeypots gather food, their sisters tidy up the mess left behind, cleaning out the entrance, readying it for the returning marauders. Within about four hours, the doorbusters come home, safe from the blazing sun, having gathered as much honeydew as they could. They go deep into the soil column where the weather's more forgiving, cooler and moister. There, they spit the nectar and honeydew they have gathered into their sisters' mouths, who store it in special stomachs (called crops) in their abdomens. Unlike regular stomachs, which have walls that let food pass into the body for digestion, crops are more like backpacks, with thick walls and valves that don't let any food out unless the workers want to let it out.

Most of the time, these receivers of sweet stuff don't release the food. They store it and hang out—literally. They dangle by their feet from the ceiling, their rumps as engorged relative to their heads as the fat end of a light bulb. As these storage ants—"repletes," as scientists call them, because they are replete with sweets—get more sweets from their sisters, their rumps get bigger and bigger until they can no longer do much except hang on. Eventually they get so large they can no longer squeeze their giant fannies through their chamber door, leaving them confined in a single, deep chamber. This cloistered madness has a reason; these rotund sisters keep the colony going in lean times.

The honeypot ants' environment does not always provide enough food. Sometimes it doesn't rain for a long time, plants dry up, and many insects, including those the honeypot ants rely upon, die out. But because of the repletes, honeypot ants can survive these hard times. When the earth grows barren, their sisters in the basement are ready to share the food stored in their giant rear ends. Workers tap those rabbit-ear antennae on their chubby sisters' heads. Their sisters respond by spitting nectar into their mouths like an all-you-can-drink soda fountain. Their remarkable bottoms, these pots filled with honey, give honeypot ants their name and help them survive when many other species can't.

Unlike the honeypot ants, my mama and I rarely come home with anything on our doorbusting list, and that's okay with us. Neither of us is very good at shopping. We mostly like watching people fight over alarm clocks and televisions, and we like going to get pancakes after. Whether or not we come home with a drastically marked-down fire pit or buy-one-get-one-free stretch pants, we'll always be able to head over to our local breakfast joint, Scratch's, for a short stack and bacon. Nobody needs to spit it into our mouths.

14 ACROBAT ANT

SPECIES NAMES: *Crematogaster coarctata* and *C. californica*

SIZE: 2.6–4.4 mm (0.1–0.18 in.)

WHERE IT LIVES: Most often, you will find acrobat ants on the forest floor, nesting at tree bases, under stones, or beneath bark or rotting wood, but sometimes they wander into our homes, snuggling their nests in tight spots like between shingles and in the walls. *Crematogaster coarctata* are found on the North, Central, and South Coasts, in the Central Valley, and in the Sierra and Coastal Mountains. *C. californica* are found on the South Coast and in the desert.

WHAT IT EATS: Primarily sugar lovers, acrobat ants sometimes take a break from lapping honeydew off aphids' rear ends to forage on protein such as dead insects.

One summer, I traveled to a remote North Carolina island for a research project. The project required that I crawl under and around people's homes looking for ants. While walking to one home, I accumulated a following of local ducks that waddled behind me, waggling their bills in the hope of getting food and quacking reproachfully when none appeared. I hate to disappoint, so I snuck into a local's backyard and dumped out my supplies, looking for a duck-suitable snack. As I rifled through my bug-collecting equipment, a man came out of the house.

"Hello, ducks ducks ducks!" he said, and the ducks happily abandoned me for their old friend. Thinking I was caught trespassing, I shoved my equipment back in my bucket and hurried to introduce the man to this potential thief/weirdo lurking around his backyard. He told me he could hear me coming but couldn't see me; he was blind.

"I'm looking for ants," I explained.

"Ants?" he asked. "I've got acrobat ants! Come see!"

The ducks and I followed him. He felt the way off his back porch, running his rough hands along the brick walls of his house, around the corner, and pushed his body behind a hedge. He pulled back branches from a wax myrtle tree and revealed a pipe leading into his house. On that pipe? A parade of acrobat ants, their little heart-shaped fannies waving in the sun!

I tried to imagine how he could find this tiny treasure so deeply hidden.

"How in the world could you tell these were acrobat ants?" I asked.

"Because," he said, and he slammed his hand down on the pipe, smashing a couple of workers. When he lifted his hand, I watched the stunned workers stumble about, smoothing their crumpled legs and antennae, gradually going back to work. "You just can't squish the jimdurn things."

He was right; acrobat ants seem to defy squishing.

A stinger, rarely used on humans, tips the acrobat ant's heart-shaped gaster.

Acrobat ants are a gift, a joy, and you can find them almost any-where you'd imagine in the United States, from swamps, deserts, and forests to your kitchen cabinet. Two species of acrobat ants make the most common California ant species list: *Crematogaster coarctata* in Northern California and *C. californica* in Southern California. These species can be hard to tell apart just by looking at them. About half the size of an apple seed, they range in color from rusty with dark brown/black abdomens to a deep reddish-black all over.

Even so, you can tell acrobat ants from other types of ants by their heart-shaped bottoms, or gasters. They trail in happy lines to and from food. When disturbed, acrobat ants halt and wave these hearts in the air like proud flag bearers in a pageant.

It's hard to imagine how acrobat ants are among the most abun-dant ants in forests, deserts, and homes, considering what a fragile enterprise colony-founding is for them. Imagine a backyard tree in which acrobat ants might live. Picture the myriad branches, stems, and leaves, jutting out against the sky. Now consider a queen, one tiny newly mated queen, slightly bigger than an apple seed, embark-ing alone for the journey of her life.

The queen must find an abandoned beetle passageway, termite gallery, or the perfect stone or unoccupied plant in which to make her new home. She must make such a discovery while avoiding spiders, mice, beetles, and birds—all of which would like a bite of her bottom. The odds are low, one in 100 or less, and yet some queens do survive. A few of those that survive thrive, building both empires and

generations. Once formed, a colony can live 10 to 15 years and may have several thousand workers crawling across the branches, eating everything from nectar to other insects.

Acrobat ants milk other insects for honeydew.

Acrobat ant workers help keep the ecosystem healthy and balanced. Across the United States, acrobat ants protect or sustain at least two endangered species: the Miami blue butterfly and the red cockaded woodpecker. In exchange for a sweet substance produced by Miami blue caterpillars, acrobat ants feistily fend off would-be butterfly poachers like birds and other ants. They also are the red cockaded woodpecker's primary diet. Wiping out acrobat ants could have a domino effect across the environment, with other species falling down in turn.

Consummate hosts, acrobat ants often harvest clytrine leaf beetle eggs from leaves and, without eating them, bring them into their nests, where the eggs hatch in a predator-free environment. Another ant-loving beetle, *Fustiger knausii*, spends most of its life hanging out in acrobat ant nests, relaxing with the brood and riding around on workers' backs. They groom the ants and might get food by enticing workers to spit up snacks for them to eat! Just what the ants get for their hard work is less clear.

Like my duck-loving friend on the island, you'll find acrobat ants parading around your kitchen or, true to their name, tightroping across your clothesline. Don't be afraid of them! They aren't dirty and they won't hurt you. Many of us commonly encounter acrobat ants and don't realize it. That's because many of us, unlike my friend, choose to be blind, to ignore these marvels of life as they shiver all around us. Maybe you, like my friend, can take a break to experience the pageantry of the happy procession before you. To enjoy the sensation of those cheery bottoms waving in the air on their way to work. To thank them for the job they do. Just try not to squish them.

FREQUENTLY ASKED QUESTIONS

Ant researchers often get asked some really interesting questions. For example, what do you do when you get ants in your pants? (Jump around and squeal, obviously!) We like that people pay attention to what's living around them and want to know more. I asked some of my favorite ant researchers what they most often are asked about ants. Here are the answers to some of our most commonly asked ant questions.

What is the biggest ant?

The dinosaur ant, *Dinoponera gigantea*, boasts the largest workers in the world, measuring a little over an inch to more than 1.5 inches long. These whoppers live in South America. But the contest is a close one. At 1.2 inches long, both the Southeast Asian giant forest ant, *Camponotus gigas*, a close relative of our US carpenter ant, and the bullet ant, *Paraponera clavata*, from Central and South America, give dinosaur ants a run for their money. In the United States, carpenter ants can measure up to a half-inch long and probably hold the distinction of the largest ants in the country.

What is the fastest ant?

No one has calculated which ant species runs the fastest, though you might judge your local species for yourself in your backyard with some ant bait and a stopwatch. Still, some ants are known to have some pretty quick moves. One type of trap-jaw ant, *Odontomachus* spp., have jaws that shut at lightning speeds—up to 145 miles per hour. They use these quick reflexes to chomp down on prey and threats alike, but they also use their snappy mouthparts to help them battle intruders or quickly exit any scene. When danger rears its ugly head, a trap-jaw ant can either bounce the botherer off its snapping jaws, tossing the intruder aside, or bounce itself off the intruder, launching itself far from danger. Trap-jaw ants can also snap their jaws down on the ground to catapult themselves into the air and away from danger.

How do I find ants?

As you probably already know, ants are super easy to find. They're all around you! It's best to look for ants on warmer days with low wind speeds, but you can look for ants any time you want.

One of the best ways to find them is to coax them into coming out into the open by offering them a snack. Some ant species love sugar while others prefer protein, so read a little about which species you're trying to find before preparing a meal for them. If you just want to see who's around, you can crumble some pecan cookies onto an index card, a piece of paper or tin foil, or any other flattish surface and wait to see who comes to the party. Pecan cookies offer protein and sugar at the same time. Many ant researchers also use canned tuna fish in oil to find their protein lovers and honey or jelly, like apple jelly, mixed with a tiny bit of water for their sugar lovers. Some researchers mix them both together to make a stinky salad that equates to many ants' dream meal.

While you're watching your ant baits, see if you can identify which species show up. Watch to see how they interact at the baits. Some species bicker with each other, while others ignore everybody completely and stick to the task at hand. Some species lay chemical trails back to their nest to get their sisters excited about this new grocery store in town, while others will carry as much home as they can in their mandibles. Still others will run away, only to return carrying their ready-to-help sisters, whom they plop down on the food. You can follow all these ants back to their nests if you have keen eyes.

Another way to look for ants is to look under things. Logs, flower pots, stones, and mulch are all great objects to peek under or pick through. I like to carefully peel back the bark on rotting wood to see who's living there, and I'm rarely disappointed. Just keep your eyes open for insects and other creatures you might not want to find, like snakes and spiders. They like living under bark and logs as much as ants do. It can be exciting to find them, but sometimes they'll startle you. Remember to replace the log, rock, or bark when you are done peeking, though, so the ants and other small wildlife can return to their business after you amble away.

If you want to take ant hunting to the next level, you can set up pitfall traps. For a pitfall trap, you'll need a little liquid dish detergent (or other liquid soap) mixed with some water (and a little rubbing alcohol, if you have it—or, even better, ethanol) and a small, disposable plastic cup. Dig a cup-size hole in the ground and place your cup in it so the lip of the cup lines up with the top of your hole. Try to get a cup with a small mouth on it so that big stuff won't fall in. Then, fill your cup halfway with your detergent mixture. Leave it in the ground for a day or so, and when you come back, you can see who fell in. You can dump your cup's contents into a white-bottomed tray or, if you don't have that, a clear dish with a white piece of paper under it so you can see everybody. If you leave the cup in the soil too long, the insects you catch will begin to rot and the smell will probably prevent you from wanting to check, so it's best to just leave it out for a day at a time.

If you want to get even deeper into ant-finding, visit your local library and check out a great ant-hunting method book like *Ants: Standard Methods for Measuring and Monitoring Biodiversity*, edited by Donat Agosti, Jonathan Majer, Leeanne Alonso, and Ted Schultz. Books like this will give you all sorts of ideas about how to find the ants you want. Such books are written for serious ant enthusiasts, who, as we all know, include kids, bankers, teachers, and anyone else who wants to seriously study the other societies all around us.

Are ants related to termites?

Ants *are* related to termites in that they're both insects, but they're not closely related to termites, and they have very different lifestyles. While ants have queens who store lots of sperm and lay eggs, termites have kings and queens who mate repeatedly over the course of the colony's life cycle.

Ants also develop by means of "complete metamorphosis," which means the eggs hatch into baby ants (larvae) that look more like fly maggots than they do ants. These larvae must pupate (often by spinning cocoons, though some ants forgo this complexity) and undergo metamorphosis before they are adult workers who look like the ants we see on our sidewalks and trees. Baby termites, on the other hand, hatch from eggs looking like miniature termites.

To tell ants apart from termites just by looking, check their waists and antennae. Ants have narrow "wasp waists," while termites look chubbier around the middle. Ants also have elbowed antennae, whereas termite antennae stick straight out.

Are ants related to wasps?

Ants are more closely related to wasps and bees than they are to termites (or most other insects). Ants, bees, and wasps are all members of the scientific order *Hymenoptera*, which means they share both physical characteristics and common ancestors.

How long do ants live?

We're not sure how long many species of ant workers live. Their lifespans often depend on the season (and whether they get stepped on, licked up by an anteater, chomped by a bird, or taken over by a zombie parasite), but they can live anywhere from a couple of months to several years. Ant researchers typically record how long ants live by how long the colony's queen can survive, since she is the colony's beating heart.

Some ant queens live long, regal lives (if a life consisting mostly of waiting to be fed and laying eggs is regal). Others don't. For example, red imported fire ant queens (*Solenopsis invicta*) can live almost seven years, depending on where they live. Winnow ant (*Aphaenogaster* spp.) queens can live up to 13 years. Some *Lasius* queens have been documented as living nearly 30 years. Other ant queens last less than a season.

Do carpenter ants eat wood?

While carpenter ants gnaw wood to build their nests, they don't actually eat it. Like humans, they just whittle while they work.

What does an ant queen look like?

Some ant queens, like Asian needle ant queens, look a lot like their workers. Others are the same color as their workers but have tremendous gasters, good for egg laying. Ant queens usually have larger eyes and bulkier thoraxes than their workers. The ant queens of some species, such as army ants, are so large and different from their workers that they look like another species entirely.

How many different ant species are there?

So far, we know of about 15,000 ant species roaming the earth, and nearly 1,000 of those species call the United States home. Nearly half of these species have not yet been named and more remain undiscovered, which is to say new ant species may well be lurking in your backyard.

HOW TO KEEP ANTS AT HOME

So You Want to Keep Ants as Pets . . .

Acrobat ants walking tightropes, carpenter ants having some serious conversations—now that you've read about all these creatures crawling around you, you might be tempted to capture some to keep around the house. Watching ants go about their business from the comfort of your home can be fun, especially when they're going about their business in a container and not just wandering around your kitchen counter.

Ant Sleuthing

Before you decide to take the leap into ant ownership, consider becoming a grade-A ant sleuth instead. Ant stalking can offer endless fun, and you'll get the chance to observe many more species

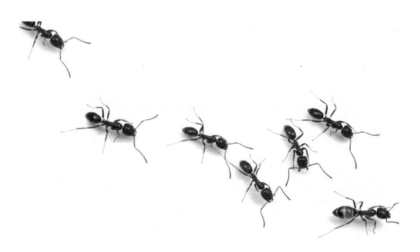

than you would if you kept a species or two in your house. To be a good ant detective, you can either lure the ants to you or hunt them down.

Luring Ants: The Way to an Ant's Heart . . .

When luring ants, consider their favorite meal. Some ants prefer sugary foods (like honeydew), some prefer protein (like arthropods), and some like a little bit of both. Start by reading about your favorite ant species to see what they like best.

Once you know their favorite fare, pick something you have around the house that offers them a taste of what they want. Pro-

tein lovers might like a little peanut butter or tuna fish in oil. Those with a sweet tooth may prefer honey, jam, or sugar water. Cookies are always a good bet. Stay away from "healthy snacks" like broccoli or celery. Ants never seem to care too much for plants (unless those plants have bugs living on them). Feel free to be an ant chef and experiment with different foods as you get to know different ant species.

You'll want a handful of mini platters so you can spread the bait out around your yard. You can use index cards, tin foil squares, or some other flat surface. Place a tiny amount (1/4 teaspoon or less) of your food on each one, and take them outdoors to see what you can find. After you've distributed their bait, wait 20 minutes to an hour and see who shows up.

Watching ants can be very exciting. Some get to the party really early and run away when other ants come around. Others saunter in and frighten away the ants already there. Some ants play well with others and ignore whoever else shows up. Others fight or engage in ritualized displays (where they wave their gasters in the air or open their mandibles or swing their heads to and fro) to show how mean they can be. I've even seen ants show up only to eat the other ants eating the baits!

See how the ants find the baits and how they bring their sisters back to get more food. Do they wander in, or do they make a direct line? Do they come alone, or do they bring reinforcements? If you place two different types of food out for the ants, do they prefer one over the other? You can come up with many questions while luring ants and get a lot of insight into how ants operate just by watching them eat. Be sure not to breathe on them while you watch them. They hate that.

Hunting Ants: No Weapons Necessary

If you've set up a bait, you have a good eye and are quick, you can follow the ants back to their nests. Many of us are not so quick, though, and we need to come up with other ways to hunt ants.

Hunting ants can seem tricky at first, but really it's not so hard. In the same way you figured out what they like to eat, you can figure out where they like to live. Again, start by doing your research. Pick a species or two that live in your area and read about where they like to make their homes. For example, odorous house ants like to live in mulch, and winnow ants like to live in rotting logs.

Next, carefully inspect those areas. On ant hunts, I carry a small but sturdy garden spade to help with delicate digging and exploring. With winnow ants, for example, I like to go into the woods and carefully peel back the bark on rotting logs. It doesn't take long before I find a winnow ant colony in there, running about with their brood in their mandibles, trying to get away from me. I find many other species that way, like acrobat ants, carpenter ants, and field ants. In mulch around the bases of trees or next to houses, I often find Argentine ants or odorous house ants, as well as the occasional tiny thief ant or field ant colony. If you have ants living in your home, try to hunt them down by following their trails until you find a nest.

Be careful while you hunt for ants. Know which ants sting and always keep an eye out for other animals that could hurt you—like snakes, centipedes, or even raccoons— that might not want you poking around their homes. More than once, I've plunged my spade in a hollow log without looking only to be greeted by a cohort of angry yellow jackets, ready to tell me they don't like what I've done to their house. If I had paid closer attention, I would have noticed the wasps coming and going from their home and avoided a painful situation for me and my ant-hunting friends. Not paying attention can make you quite unpopular with your companions.

Keeping Ants as Pets

If watching ants in their natural environment isn't enough, and you still want to try your hand at keeping them at home, you have a number of ant-keeping options at your disposal. To get a better idea of what sort of housing might be appropriate, think about the type of ant you'd like to have and its preferred lifestyle. Ant houses work well for species that thrive in enclosed environments, but not others that like to forage over larger areas or up and down tree trunks.

Selecting Your Ants

As you know from reading this book, the world around you is bursting with ant species. Each has its own unique way of life, including diet and housing. You can purchase ants online from one of many ant retailers, but be warned that most retailers send several workers and no queen. Without the queen, the colony will not be able to replenish workers when they die, and the colony will die out in a few months to a year or so. Still, purchasing workers online can be an easy ant-acquiring solution and a rewarding experience for beginner ant keepers.

If you choose to capture your own ants, consider:

× which ant species live around you and are active when you want to collect them

× where those species live (in logs, underground, in mulch, under rocks, high in trees)

× what those species eat (some diets may be more difficult to reproduce at home than others)

× if those species are dangerous and put you or your household at risk. (You will probably want to avoid collecting Asian needle ants or red imported fire ants, for example, as they sting.)

× if those species could become household pests. (When I studied ants in a laboratory, I accidentally infested my laboratory with Argentine ants more than once. Others do not appreciate this.)

× if the species you select are invasive. Releasing invasive species into the environment, even accidentally, is illegal and can be harmful to the environment

Now that you've decided who you want to bring home, it's time to build or purchase your species' dream house. We'll collect the ants later, when we're ready.

Home, Sweet Home

Ant houses, called formicaria, can be purchased from a number of online retailers. Formicaria constructed from natural materials and customized for a particular ant species work best to ensure you'll be able to keep your ants for a long time. For those just getting started, I'm a fan of purchasing a good formicarium over building one at home. That way, somebody will mail to your house everything your ants need to get rolling, and you can see what works for others before you try it yourself.

When building an ant house, consider how the ant lives outdoors.

If you do decide to build your own ant house, remember that many species can escape containers easily, so you need to have containers with tight-fitting lids. These lids should be fitted with fine screens or have tiny ant-proof holes for proper ventilation.

Formicaria can take many different, beautiful forms.

Ants also need proper moisture. The simplest ant nest can be built using a test tube, a stopper with a hole in it large enough for the ants to crawl out, cotton, tin foil, and water.

Fill the test tube about three-quarters of the way with water. Stuff cotton in the tube with clean fingers until you feel the cotton moisten to your fingertip. You don't want the tube to leak, but you do want it to be moist. Then, put the stopper in the tube. Be sure there's enough room in the tube for your ants to crawl around. Wrap the tube in tin foil so the ants can have a dark environment. You can peel the foil back to observe your colony from time to time. This tube should be placed in your container with tight-fitting lid.

You can feed your ants by placing their food directly in the container. Check your test tube every week or so to ensure it stays moist. If it dries, add another test-tube nest. The ants will move in when they run out of water, and then you can remove their old nest.

Remember, ants do not like disturbances, so with a homemade ant house like this, you should only check on them once or twice a day at most. When you check on them, try to avoid breathing on them, as the carbon dioxide in your breath alarms them.

After you master this basic ant setup, you may want to modify your homemade formicarium to impress your friends, suit your ant-watching desires, or help your ants to live in a more "natural" environment.

If you want to build a home where you can watch your ants crawl around all day, check the Internet for some sample formicaria plans, which will provide you with step-by-step instructions on how to build your ant's dream home. You can also play around with expanding upon your existing setup.

Go Get Your Ants!

Although I do recommend purchasing a premade formicarium, I also recommend collecting the ants for that formicarium yourself. When you collect your own ants, you can get as many as you want, and you can ensure you have a mated queen that could potentially provide you with years of ant-watching pleasure.

The simplest way to collect ants is to go on an ant hunt, outlined above, armed with your spade and containers with tight-fitting lids. When you find the species you want, try to locate the queen, some workers, and a bit of brood. Queens are usually larger than the other ants in the nests and can often (but not always) be found near the brood. Delicately scoop up the queen, workers, and brood with a spade and place them in your container with a tight-fitting lid.

When you get home, place your test-tube nest in that container and wait for the queen and workers to move in. Leave them alone for a day or two so they feel comfortable.

Supper Time!

Now that you have a small colony, it's time to give them something to eat. Most ant species need protein, sugar, fats, and water. Read up on your species to see what types of food they like most. As you work to make fatties, keep in mind that just because they prefer sugary foods doesn't mean they don't need a little protein and fat every now and then. You can purchase ant food online, or you can give it a go yourself. Here is a list of foods I use:

PROTEIN
- ✕ Dead arthropods (sized appropriately for your species)
- ✕ Peanut butter (works sometimes but not all times)
- ✕ Seeds and nuts (for ants interested in seeds and nuts)
- ✕ Tuna fish in oil (they love it, but it smells bad)
- ✕ Pecan cookies (makes a nice snack but will not sustain the colony)

SUGAR
- ✕ Sugar water (50 percent sugar, 50 percent water)
- ✕ Honey mixed with a little water
- ✕ Apple jelly or some other fruit jelly mixed with a little water

FAT
- ✕ Tuna in oil has fats, as do seeds and peanut butter (above)
- ✕ Some people mix a little olive or canola oil with their sugar solutions

With all of these foods, it's important to remember ants' tiny stomachs. Because you want to avoid getting mold in your ant houses, try not to feed them too much, and remove uneaten food as it dries out or begins to mold. I like to feed them on tiny trays I fashion out of tin foil or in shallow bottle caps, so I can remove uneaten food easily.

Watching Your Ants

Now that you have ants in your living room, spend some time watching them to see what they do. How do they eat? When you put out new food for them, how do they find it? How long does it take them to find it? Which foods do they pick up most often and most readily? Which foods do they avoid? What do they do in their nest chambers all day? Does light bother them? What times of year does the queen lay eggs? How do they feed their larvae? Watching ants can be fascinating. You'll be amazed how quickly the time goes.

PRO TIPS FROM A PROFESSIONAL ANT COLLECTOR
Keeping ants at home can be complicated. Mack Pridgen, founder of the ant house business Tar Heel Ants, has figured out some valuable tricks to the trade, and he wants to share them with you.

DR. ELEANOR: How did you get interested in keeping ants?
MACK: I was fascinated by ants as a kid. We had some very large colonies of carpenter ants (*Camponotus*) around my house, and I was the kid with the jars trying to collect them and feed them. My favorites were the giant red ants with big heads! I now know those as the carpenter ant *Camponotus castaneus* majors.

Years later I bought my daughter a gel ant farm as a Christmas gift and decided to show her how cool ants were to watch digging. Instead of ordering ants (when I was a kid, mine arrived dead), I went out the day after Christmas and found carpenter ant *Camponotus chromaiodes* workers foraging up a tree. They died within a couple of days. I pressed on, researching how people cultivate ants in labs, and I reached out to the local ant lab here in town. Before long I was raising young colonies in test tubes and planning their first formicarium!

DR. ELEANOR: What common ants would you say are the "best" (easiest, safest) for beginners to keep at home?

MACK: It depends on where you live. Concentrated in the southeastern United States but also found in other regions throughout the country, various winnow ants are abundant in forests under rocks. Winnow ants are always my first suggestion. I was lucky enough to have them for my first large colony.

Carpenter ants (*Camponotus* spp.) also work well for beginners. Their queens are huge, around three-quarters of an inch in size, and they vary in color.

DR. ELEANOR: What do you find to be some of the common mistakes people make when keeping ants at home?

MACK: Common mistakes include

1. **Improper feeding expectations and procedures.** For example, most queens do not need food until their first workers have emerged. The workers find food and feed the queen. Many inexperienced ant keepers try to feed ants live insects, too much of a dead insect, or sweet liquids inside their claustral chambers (test tubes or other starting formicaria). This stresses the queen, causes mold problems for the queen and her brood, and more. Carefully feeding small amounts of foods is OK but the excess should be removed immediately. Practicing a proper feeding regimen can make all the difference for your colony, and an improper regimen is often an ant keeper's biggest problem. Follow these proper feeding techniques:

   × Offer small pieces of fresh insects often—at least a couple times a week—and anchor them to something (cork, silicon, etc.) to prevent trash from entering the formicarium (less important if you can disassemble your formicarium).
   × Always provide liquids and solids on dishes, regardless of colony

size. Any food that smears on the formicarium surfaces or foraging area can mold and lead to major problems for the colony down the road. Providing some loose (ant-safe) substrate allows your ants to cover the problem areas themselves.

× Never feed your colonies live food. Workers can get injured battling living insects. At minimum, injure prey if live feeding is a must for your particular species.

× Rotate foods. Use a combination of fresh fruits (apples are a personal favorite), dead crickets, mealworms, and fruit flies, in addition to other foods that provide a good balance of carbohydrates, proteins, and fats.

2. **Not anticipating how easily ants can climb and escape.** One of the first major considerations while planning your homemade formicarium is security. Think of this as protection for your ants, not you!

3. **Not planning ahead.** Colonies can grow fast during their growth season. Many of the species we see in urban areas, such as pavement ants, fire ants, and acrobat ants, can quickly outgrow their formicaria and can be difficult to contain and feed if not kept in a proper-size habitat.

DR. ELEANOR: What are your top tips for building an ant house at home?

MACK:

1. **Use what you already know.** You can build an ant home using skills you already have. Good with crafts? Perhaps build a plastic nest out of acrylic or make a plaster nest. Artistically inclined and like to carve? Good with power tools? Try finding some Ytong (AAC block) bricks and crafting your own design.

2. **Plan your hydration first.** Moisture is your colony's lifeline. If you can, use a hygrometer to test the moisture level in your formicarium before introducing your ants.

3. **Ask questions.** Many people have been where you are now. Join an online group or forum.

4. **Keep local ants!** Don't risk spreading harmful species.

DR. ELEANOR: What are your top five tips for keeping ants alive at home?
MACK:

5. **Monitor the temperature** near your colonies daily with a thermometer. Keep your ants at a constant temperature (78–82°F, typically, though some species may prefer warmer conditions). Quick temperature changes can be problematic for ants. Remember: Ants are not plants. They don't do well in the sun!
6. **Label your ant colonies** in case friends or families don't know what they are. Shaking or tilting your formicarium is not recommended.
7. **Wear disposable latex gloves** or, at a minimum, wash your hands before feeding your ants or handling their formicaria.
8. **Use organic fruits and honey** when possible.
9. **Clean out your ant foraging areas regularly**, whenever you see detritus build up.

EPILOGUE

At some point during the Cretaceous period, when dinosaurs dominated the land, the first ants walked the Earth. We know little about the habits of these early ants, but it appears they were not abundant and did not form large colonies. Over the last 100 million years, these early ants gave rise to one of the most ecologically dominant animal lineages on the planet today. To some people ants may seem invisible and insignificant. In actuality, ants rule the world as measured by sheer biomass and ecological importance. Though they may be small in stature, they have a huge impact on ecosystems. And what they lack in size, they more than make up for in numbers. These ubiquitous insects are also the most behaviorally diverse social animals alive today. I call ants the glue that holds ecosystems together.

I remember meeting a *Formica integroides* colony for the first time in the Santa Cruz Mountains. Normally you don't expect to find such a large mound building ant in the Bay Area, but there they were, swarming over a large mound more than three feet across, marching along four trunk trails leading to foraging areas out in the forest. It was like finding a magnificent elephant living right here in the Bay Area. I stood and watched one of the trunk highways, with three rows of ants heading out and another four lanes of traffic returning with a bounty of freshly harvested insects. It seemed as if the ants were pulling all insect life from the forest. Some were carrying large green caterpillars while others were carrying just parts of larger insects. When first looking into the face of a returning worker, one can only be grateful that these ants are not the size of dogs with jaws set sharply around their prey. Having just navigated the Bay Area traffic to the Santa Cruz Mountains, I was also amazed by how much better they are than humans at engineering traffic. There was no gridlock or slowdown as traffic increased.

Some ants that passed held no food, but their abdomens were swollen nonetheless. These ants had been tending aphids or other insects up in the vegetation. Ants milk these insects for sugar-rich plant phloem, which the insects excrete as honeydew. The ants hold the liquid store in their crops, or "social stomachs" to share with others back in the nest. A large portion of the colony's energy comes from such symbiotic relationships with aphids or other hemipterans.

If I were an ant, I wondered, would I be altruistic enough to take a juicy caterpillar back to the nest? After all, it was almost noon and I could have used a snack. Maybe, I thought, I would just sneak behind a rock to eat the caterpillar, or maybe just half? Actually, if I were an ant, I wouldn't have a choice—adult ants can't eat solid food. They have to take the food back to the nest because that is where the colony's stomach is located. The larvae in the colony are fed the solid food, digest it, and then regurgitate the nourishment in liquid form to be shared throughout the colony. This illustrates the point that to understand ants, you must look at them as a superorganism. The organism is not any individual worker ant, but the colony as a whole.

Those who study ants are called myrmecologists. One of the first steps involved in studying ants, and the focus of my research, is to discover what ants live where. Museums like the California Academy of Sciences in San Francisco play a special role in documenting the species diversity and distribution of this "invisible majority" of insects. We still have a long way to go—we estimate that only half of the world's species of ants have been discovered. In the Bay Area and Delta regions of California, there are more than 100 ant species, while there are almost 300 species that occur in our state. All of them have much to teach us about how they live and why they are found where they are.

As a myrmecologist, I dream of the day when we can easily find out the name of any ant; when we can learn not just its name, but what it looks like, its habits, its distribution, whether it is endangered, whether it is invasive, and whether it stings. We could use this knowledge to help protect habitats, monitor ecosystems, and link the health and well-being of humans to that of the natural world.

To get there, we are enlisting the help of curious minds across California. These people—interested in the wild things that are their neighbors—are helping us discover, document, and sustain the ants and their ecosystems. By enabling local experts to participate, we will move this ubiquitous group of animals toward an equal footing with birds in monitoring the profusion of life we know as biodiversity and help ensure they continue to flourish alongside humankind. Keeping an eye on insects is also good for the human spirit. As Zana Briski reminds us in her work with praying mantises, honoring insects inspires reverence. Maybe that is why His Holiness the Dalai Lama advises: What is the most important thing to teach our Children? Teach them to love the insects.

ACKNOWLEDGMENTS

This book would not have been possible without the expertise and help of many great individuals. Thank you, Rob Dunn and Andrea Lucky, for having the curiosity and vision to help us all turn over stones and peer in the grass to find the ants around us. Rob, thank you, also, for your editorial guidance, opportunities, and encouragement. You and the University of Chicago Press have opened the door for us to share our joy of ants. My great appreciation goes to the University of Chicago Press's Christie Henry, Gina Wadas, and Amy Krynak for their guidance and support. Robin Anders, you know how to edit and critique like nobody else. Thank you. Thank you also, Kathryn and Jamesie Spicer, for your editorial assistance and for giving me the crumbs and long mornings and afternoons to meet my ants. Neil McCoy, thank you for your creativity, which helps to sharpen and enliven ideas. Holly Menninger, your powers of coordination and organization are unparalleled. Thank you, Alex Wild, for using your skill and vision to make giants of ants, for showing us how beautiful their tiny world is and how they can be our friends. Thank you, Brian Fisher, for sharing your vast experience with us, and writing the book I carry with me whenever I go looking for ants. Thank you for your expert advice and encouragement, Matt Shipman. Russ Campbell and the Burroughs Wellcome Fund, thank you for recognizing this book and giving it a chance. Thank you to my ant people, including Jules Silverman, Alexei Rowles, John Brightwell, Bill Reynolds, Sean Menke, Jon Shik, Brad Powell, David Bednar, Grzesiek Buczkowski, Heike Meissner, Benoit Guenard, and Clint Penick, plus one termite guy, Warren Booth. Greg Rice, you are the words and the ants and everything else. You make everything possible for me. Thank you.

GLOSSARY

abdomen: The third major division of the insect body (aka rump, booty, posterior, etc.) that contains most of the ant's organs and its stinger.

ant: A small, wingless, wasp-like insect that usually lives in eusocial groups. An incredibly diverse and ecologically important animal. Earth houses more than 15,000 named ant species, and many more awaiting scientific description and naming. *See* Hymenoptera and Formicidae.

antenna (pl. antennae): A segmented appendage projecting from either side of an adult insect's head. Antennae function as sensory organs and help ants sniff, feel, and taste.

aphid: A small, plant fluid–sucking insect that usually resembles a tiny cicada or a tiny, chubby katydid. Aphids can be winged or wingless and usually are found on the undersides of plant leaves or along stems. Often protected by many ant species, aphids turn excess plant fluid into a sweet substance called honeydew, which ants eat.

arthropod: An animal with jointed legs and an exoskeleton. Arthropoda refers to the large scientific group including shellfish, insects, scorpions, and spiders. *Arthro* comes from the Greek word meaning "joint," and *poda* comes from the Greek word meaning "foot." The vast majority of described species on earth are arthropods. Ants are arthropods.

biodiversity: The amount of different life forms in an area. In general, a rich biodiversity (lots of different life forms) means a healthy environment. Some invasive ant species, like Asian needle ants, reduce biodiversity when they move into an area, which could result in an unhealthy habitat.

caste: Refers to the various groups of ants within a colony. Sexual castes consist of two groups: males and females. Morphological castes consist of two or more groups, typically minors and majors (soldiers). Temporal castes divide ants according to age and the jobs they do at those ages. Reproductive castes refer to queens, which reproduce, and workers, which don't.

colony: A group of ants, often closely genetically related, which operate as a functional unit without aggression among the group. Colonies can have one or many nests and one or many queens.

common name: The moniker we call ants for convenience. Most people referring to ants use their common names. Common names usually refer to some aspect of the ant's appearance (like "little black ant") or behavior (like "fire ant"). Common names can be different in different languages.

complete metamorphosis: A form of insect development, in which the insect undergoes the following stages to adulthood: egg larva (looks very different from adults) pupa adult. Ants undergo complete metamorphosis.

crop: A "stomach" attached to the esophagus that serves to receive and hold food. It's like an internal backpack. Crops hold food without digesting it so ants can share it with their sisters or eat it later.

ecology: The study of the relationships between living things and their environment. *Eco* comes from the Greek word for "house," and *ology* comes from the Greek word meaning "the study of."

egg: The first stage in an ant's development and laid by queens, an egg has a simple germ cell, nutritious yolk, and a surrounding membrane.

entomology: The study of insects and other arthropods. *Entom* comes from the Greek word for "insect," and *ology* comes from the Greek word meaning "the science of."

entomologist: Someone who studies insects and other arthropods.

eusocial: If an animal cooperatively cares for its young, has a reproductive division of labor (for example, queens reproduce; workers work), and an overlap of at least two generations sharing a space and contributing to the group, then it's eusocial. Most ant species are eusocial. The few that are not eusocial are workerless parasites in the nests of other ants.

exoskeleton: The "hard outer shell" of insects and other arthropods. Instead of bones, insects have a suit of armor consisting of a plastic-like substance called chitin, which is covered by a thin layer of waxy material. Ant muscles are attached on the inside of the exoskeleton.

exotic: In invasion ecology, exotic refers to an organism that is present in an area but which comes from a different place. That is, that organism did not evolve in that area. Exotic species are not always invasive, and they're not always pests. Your pet cat is an exotic species, and some would say it is a pest. Honey bees are also exotic species in the United States. They come from Europe and Africa.

Formicidae: The scientific grouping called family to which all ants belong. The word *Formicidae* comes from the Latin word meaning "ant."

gaster: The swollen part of the abdomen behind the ant's skinny waist, or petiole.

genus (pl. genera): A group of species that share characteristics and are often closely related. For example, thief ants and red imported fire ants share many physical characteristics and are closely related. They share the genus *Solenopsis.* Knowing genera can help you mentally group ants by form and function.

holometabolous: The quality of an organism, like an ant, of undergoing complete metamorphosis.

honeydew: A sugary fluid excreted from the abdomens of many different insects, including aphids and scale insects. Many ant species love to eat honeydew and rely on it for survival.

Hymenoptera: The scientific order of insects to which ants belong. Bees and wasps also belong to this order, and these three types of insects share much in common, including their skinny waists and their tendency toward forming social groups. Hymenoptera comes from *hymen,* the Greek word for "membrane," and *ptera,* the Greek word for "wings." Hymenopteran wings look like a thin membrane stretched across a few veins.

insect: A class of animal that has an exoskeleton, three major body segments (head, thorax, and abdomen), six legs, and two antennae. Ants are insects. Spiders (eight legs, two segments) are not.

invasive species: A species that moves into an area and negatively impacts that environment. Red imported fire ants and Asian needle ants are examples of invasive species.

invertebrate: A general term referring to any animal that does not have a backbone. Worms, insects, crabs, octopi, and spiders are all invertebrates. Most of life on earth has no backbone.

larva (pl. larvae): The second stage in an ant's development, between egg and pupa. Larvae differ in form from adults. Ant larvae often look like legless grubs.

major: A worker subcaste (*see morph*) in which the individual is typically larger and specialized for defense. Big headed ants have the most prominent majors, but other ant species, like carpenter ants, can have majors, too. Sometimes referred to as soldiers.

mandible: The first pair of jaws in ants. Mandibles usually stick out from the front of the head and are good for chomping, slicing, and carrying.

minor: A worker subcaste (*see morph*) in which the individual is typically smaller and specialized for work.

morph: Any of the various forms of ants within a caste. For example, a major is one morph, while a minor is another.

myrmecologist: A person who studies ants.

native species: An organism that is present in an environment "naturally" and not because a human facilitated its presence in the environment.

nest: Among ant species, a nest is a discrete living space for a related group, usually containing workers, brood, and queens, but sometimes containing any two of the three (or any combination plus males). Nests can be as simple as a hangout spot under a rock (as with odorous house ants) or as complex as intricate underground tunnel networks connecting rooms (as with winter ants). Ant species can have one or many nests per colony.

nestmate: Individuals, usually related, who share a nest. Nestmate can also refer to members of the same multinest colony who don't share a particular nest. Nestmates do not fight one another when they meet outside the nest, and they recognize one another as nestmates because they smell alike.

pest: A species that negatively impacts its environment. Some ant pests, like odorous house ants, are "nuisance pests," meaning that people tend to be bothered by them but they don't necessarily negatively impact the environment. The primary damage associated with these pests is indirect, resulting from people's use of hazardous chemicals to exterminate them. Other pest ants, like red imported fire ants, are true pests, meaning they cause economic damage (like crop loss), or pose health risks (as from the sting of an red imported fire ant). Not all pests are invasive species or exotic species, and not all exotic species are pests.

petiole: The skinny segments at the beginning of the abdomen, between the thorax and gaster, that give ants their skinny, wasplike "waists."

pheromone: Any one of many chemical secretions used to communicate within species. Ants use a variety of pheromones to communicate, including alarm pheromones, recognition pheromones, and trail pheromones.

polyethism: The division of labor among members in the colony. Different forms of polyethism are apparent in ant colonies. For example, many ants display something called age-based polyethism, where younger workers perform different tasks than older workers.

polymorphism: In ants, having several physical forms of adults. Many insects display polymorphism.

pupa (pl. pupae): The life cycle stage in insects with complete meta-morphosis. In ants, it occurs between the larva and adult stages, when the insect becomes inactive, doesn't eat, and develops the physical features of an adult.

queen: In ants, female colony members who can lay fertilized eggs. Usually larger than workers.

scale insect: A small, plant fluid–feeding insect that looks like a bump, shell, or scale stuck to plant bark or stems. Often protected by many ant species, scale insects turn excess plant fluid into a sweet substance called honeydew, which ants eat.

scientific name: The formal epithet used to describe species; regu-lated by a huge international formal naming process. Usually with Greek or Latin roots, scientific names are the same in all languages across the globe. This standardization is extremely useful for communicating science. Just like we have first and last names, scientific names consist of two parts: one for genus and the other for species. As knowing somebody's name can tell you about that person, knowing scientific names can tell you a lot about the insect. The genus name is like our last name and the species name is like our first name. For example, my name is Eleanor Spicer Rice. If I told somebody from my hometown my name, that per-son would know I'm kin to the Spicers and could have a general knowledge about me before they even got to know me. If she knew my relatives, she could get an idea of what I might look like and could have an idea of where I live and to a certain degree how I might behave. If you tell an ant scientist you saw a *Pachycondyla chinensis*, even if he's never met one, he would know a lot about how the species looks, lives, and acts if he knows other *Pachycon-dyla*. Just as my first name, Eleanor, distinguishes me from the other Spicers hanging around town, the specific epithet distin-guishes each species from all the other species and gives us an idea of what that species does. For our *Pachycondyla chinensis*, "chin-ensis" tells us this species is native to Asia. While species might have the same specific name, no two species share both genus and species name. That way, there's no confusion about which species scientists are talking about.

segment: In insects, any division of the body. While segment can refer to each joint in the leg or antenna, we most often think of segments when discussing one of the three major insect body divisions: head, thorax, and abdomen.

soldier: See major.

species: A group of individuals that are genetically similar and able to mate and produce offspring that can also mate and produce offspring.

spiracle: The holes on an insect's body that open to its respiratory, or tracheal, system. Basically, it's how the insect breathes. Like our mouth or nose.

thorax: The second, or middle, segment of an insect. The thorax is the locomotion center.

trophallaxis: In ants and other eusocial insects, the process of exchanging crop contents between individuals through the mouth. It's one way ants share food and communicate information.

worker: In social insects like ants, a member of the laboring caste that isn't able to reproduce.

ADDITIONAL RESOURCES

Throughout the book, we relied on a collection of excellent resources that we hope will help you, too, as you continue your love affair with ants.

WEBSITES

× Antweb (www.antweb.org) offers the world's largest ant-centric database, complete with photographs, cool ant stats, and the latest research from curators across the country.

× Alex Wild snaps up-close photos of thousands of ants and displays them on www.alexanderwild.com.

× Joe MacGown's beautifully illustrated ant keys may help you identify your formicid neighbors. Explore his findings, plus a bunch of other great insect information, on the Mississippi Entomological Museum website (http://mississippientomologicalmuseum.org.msstate.edu).

× With AntWiki (www.antwiki.org) and AntCat (www.antcat.org), you can check out all the ant taxonomic information you could ever want.

× And if you ever want to read scientific literature about ants, a great place to start is the USDA's FORMIS page (http://www.ars.usda.gov/News/docs.htm?docid=10003).

BOOKS

× *Ants of North America: A Guide to the Genera*, by Brian L. Fisher and Stephan P. Cover, is the best ant field guide around. In addition to the helpful and beautiful ant graphics, the authors give you an engaging natural history of each genus, helping you to get to know these ants on a deeper level.

× *Journey to the Ants. A Story of Scientific Exploration*, by Bert Hölldobler and E. O. Wilson is an exciting, beautiful book about the discovery and love of our natural world (through the eyes of two of the world's best and most treasured ant lovers).

- × *Adventures among Ants. A Global Safari with a Cast of Trillions*, by Mark Moffett. The "Indiana Jones of Ants" shares personal stories and interesting ant behaviors around the world in this page turner.
- × *The Ants*, by Bert Hölldobler and Edward O. Wilson, is the definitive guide to all things ants. Despite its intimidating size, this book is an engaging read and illuminates every ant nook and cranny, from ant evolution to their array of behaviors to their intricate physiologies and taxonomic details.